Thermal Hydraulics

Volume I

Authors

Maurizio Cumo
Professor
Department of Energetics
University of Rome
Rome, Italy

Antonio Naviglio
Professor
Department of Energetics
University of Rome
Rome, Italy

CRC Press
Taylor & Francis Group
Boca Raton London New York

CRC Press is an imprint of the
Taylor & Francis Group, an **informa** business

First published 1988 by CRC Press
Taylor & Francis Group
6000 Broken Sound Parkway NW, Suite 300
Boca Raton, FL 33487-2742

Reissued 2018 by CRC Press

Library of Congress Cataloging-in-Publication Data

Cumo, Maurizio.
Thermal hydraulics / authors, Maurizio Cumo, Antonio Naviglio.
p. cm.
Includes bibliographies and index.
ISBN 0-8493-6787-5 (set). ISBN 0-8493-6789-1 (v. 1). ISBN 0-8493-6790-5 (v. 2)
1. Heat—Transmission—Data processing. 2. Fluid dynamics—Data processing. I. Naviglio, Antonio. II. Title.
TJ260.C86 1988 87-15524
621.402' 1—dc19

A Library of Congress record exists under LC control number: 90002055

Publisher's Note
The publisher has gone to great lengths to ensure the quality of this reprint but points out that some imperfections in the original copies may be apparent.

Disclaimer
The publisher has made every effort to trace copyright holders and welcomes correspondence from those they have been unable to contact.

ISBN 13: 978-1-315-89816-2 (hbk)
ISBN 13: 978-1-351-07726-2 (ebk)

Visit the Taylor & Francis Web site at http://www.taylorandfrancis.com and the
CRC Press Web site at http://www.crcpress.com

PREFACE

The development and diffusion of computer programs and design codes in the last several decades has allowed the design and analysis of complex fluid systems and of "advanced" heat transfer equipment. Nuclear reactor plants and some chemical plants require (both at a design and at a safety-analysis stage) analyses in transient conditions; the chemical industry requires more and more advanced solutions for heat transfer components; cryogenic systems impose the solution of complex thermomechanical problems; the aircraft industry has in heat transfer performances one of the greatest limits to the development of its components.

This text, including the description of the most relevant phenomenologies and of some advanced techniques in heat transfer with fluids, is mainly aimed at engineers using design or computer analysis programs and codes, in order to achieve a deeper understanding of the phenomenologies and of the applied analysis methods. For the same reasons, this text may be helpful to people engaged in developing original computer programs or design methods, because they may find in it basic information on the computer program-oriented solutions of the conservation equations and of the various flow and heat transfer mechanisms. The selection of up-to-date correlations in various heat and mass transfer branches represents, for the designers using traditional techniques, a helpful instrument to integrate the basic handbooks.

Finally, the trial of representing phenomenologies and problems through elementary concepts makes this text useful to students at the graduate level involved in the study of fluid flow and heat transfer.

The limits imposed by editorial needs make the treatment itself quite concise. We, therefore, suggest the use of the volumes as a reference to the basic thermodynamic and heat transfer books, while, for deeper analyses, we suggest the reader consult the updated, selected bibliography.

THE AUTHORS

Maurizio Cumo is Professor of Nuclear Plants at the University La Sapienza of Rome, where he is also director of the Postgraduate School for Nuclear Safety and Radioprotection.

He received his doctorate in Nuclear Engineering at the Politechnic University of Milan in 1962. Since then, he has participated in much thermohydraulic research at the ENEA Research Center Casaccia near Rome as well as at the University of Rome. In this field he has authored or coauthored more than 150 scientific publications as well as two books.

Prof. Cumo is a member of the Assembly for International Heat Transfer Conferences, the Executive Committee of the International Center for Heat and Mass Transfer, the EUROTHERM Committee, and the European Two-Phase Flow Group with particular attention to experimental development. Presently, he is Acting President of the Italian Commission for Nuclear Safety and Health Protection and a member of the Board of the Italian State Agency for Nuclear and Alternative Energies (ENEA). He is also Chairman of the Italian Association of Nuclear Engineering (ANDIN), Vice-Chairman of the Italian Society of Standards (UNI), member of the board of directors of the International Solar Energy Society (ISES), and a member of the American Nuclear Society, the American Institute of Chemical Engineers, and the New York Academy of Sciences.

Biographical references are provided by Who's Who in the World, Who's Who in Europe, International Book of Honor, International Who's Who in Education, Dictionary of International Biography, World Nuclear/World Energy Directory, and the International Directory of Distinguished Leadership.

Antonio Naviglio is Professor of Thermal Hydraulics at the Department of Energetics, Faculty of Engineering, University of Rome, Italy. He received the Italian "Laurea" in Nuclear Engineering in 1973 from the University of Rome.

From 1973 to 1975 he worked as a process engineer in a major Italian engineering company. In 1975—1976 he worked for the Italian Agency for Nuclear Safety and Radiological Protection (ENEA-DISP). From 1976 to 1981 he worked as a process engineer, mainly in the field of thermal hydraulics, for the Italian Electric Power Authority (ENEL). Since 1981, he has been working at the University of Rome, first as Assistant Professor and then as Professor of Thermal Hydraulics.

Prof. Naviglio is an expert member of the Italian Committee for Nuclear Safety and Radiological Protection, the executive committee of ANDIN, the Italian Association of Nuclear Engineers and of ANIAI (the Italian Association of Architects and Engineers), and Director of UNITAR/UNDP Centre on Small Energy Resources, Rome.

Prof. Naviglio has authored or coauthored some 90 scientific publications in the field of heat transfer and energy exploitation.

The research activity of Prof. Naviglio has been mainly devoted to heat transfer phenomena, complex thermal hydraulic phenomenologies affecting the performance and safety of equipment both for nuclear and for chemical plants, and the development of innovative processes which save energy and minimize environmental impact.

ACKNOWLEDGMENTS

We are indebted to the following friends who have largely contributed to the reduction of the present text: Marco Caira, Gianfranco Caruso, Enrico Cipollone, Giovanni Elvio Farello, and Augusto Porzi.

to Gabriella and Daniela

NOMENCLATURE

a	Absorption coefficient (−)
a	Acceleration (m/sec²)
a	Sonic velocity (m/sec)
A	Flow area, heat transfer area (m²)
A_t	Area of a finned pipe, without fins (m²)
A_f	Finned area (m²)
a_1, a_2, a_3	Numerical constants (−)
B	Fin thickness (m)
b	Perimeter (m)
b_1, b_2, b_3	Numerical constants (−)
c, c_p, c_v	Specific heat, isobaric specific heat, specific heat at constant volume (J/kg K)
C	Molar concentration (mol/m³)
c_1, c_2, c_3	Numerical constants (−)
d	Prefix indicating differential (−)
d, d_i, d_o, d_{eq}	Diameter, internal diameter, external diameter, equivalent diameter (m)
d_f, d_r	Fin diameter, fin root diameter (m)
d_w, d_h	Wetted, heated diameter (m)
D	Diffusion coefficient
e, exp	Natural logarithm base and exponential, exponential (−)
e	Specific energy of a fluid (J/kg)
e	Emissivity (−)
E	Flow area reduction factor in a lattice (−)
F	Force (kg/m sec²)
f	Fanning friction factor (−)
f	Frequency (1/sec)
f	Function symbol (−)
f	Condensation efficiency (−)
G	Mass velocity (kg/m² sec)
g	Acceleration due to gravity (9.807 m/sec²)
h	Heat transfer coefficient (W/m² K)
i	Specific enthalpy (J/kg)
J	Superficial velocity (m/sec)
k	Thermal conductivity (J/m sec K)
L	Length (m), work (J)
l	Fin eight (m)
M	Mass (kg)
M	Molecular weight (kg/mol)
N	Generic number (−)
p, P	Pressure (Pa)
P, P_t, P_l, P_c, P_r	Pitch, transversal, longitudinal, circumferential, radial pitch (m)
P_w	Wetted perimeter (m)
q	Heat transfer rate (W)
q''	Heat flux (W/m²)
R_c	Curvature radius (m)
R	Radius (m)
r	Current radius (m)
r_b	Bubble radius (m)

s	Thickness (m)
S	Slip ratio (–)
S	Cross section (m^2)
s	Specific entropy (J/kg K)
s	Fin pitch (m)
t	Time (s)
T	Temperature (K), (°C)
u	Specific internal energy (J/kg)
u, v_x	Velocity in x-axis direction (m/sec)
U	Overall heat-transfer coefficient (W/m^2 K)
V	Volume (m^3)
v	Specific volume (m^3/kg)
v	Velocity, velocity in y-axis direction (m/sec)
W	Power (W)
w	Clearance (m)
w	Velocity in z-axis direction (m/sec)
x	Coordinate (m)
X	Quality (–)
X	Molar fraction of a substance (–)
y	Coordinate (m)
z	Coordinate (m)

GREEK SYMBOLS

α	Void fraction (–)
α	Thermal diffusivity (m^2/sec)
β	Volumetric quality (–)
β	Expansion coefficient (1/K)
$\dot{\gamma}$	Shear rate (1/sec)
γ	Cp/Cv ratio (–)
γ	Shear angle (–)
Γ	Mass flow rate (kg/sec)
δ	Damping coefficient (–)
δ	Film thickness (m)
Δ	Prefix indicating finite difference (–)
ϵ	Efficiency (–)
ϵ	Rugosity (absolute) (m)
ϵ	Emissivity (–)
η	Pressure critical ratio (–)
η_f	Fin efficiency (–)
θ	Nondimensional time (–)
θ	Angle (–)
θ	Nondimensional temperature (–)
λ	Latent heat of vaporization (J/kg)
μ	Viscosity (kg/m sec)
ν	Kinematic viscosity (m^2/sec)
π	Nondimensional pressure (–)
ρ	Density (kg/m^3)
σ	Surface tension (kg/sec^2)
σ_{SB}	Stefan-Boltzmann constant (W/m^2 K^4)
τ	Shear stress (N/m^2)
ϕ	Two-phase multiplier (–)

SUBSCRIPTS

a	Absolute, annular
acc	Acceleration
av	Average
b	Bubbles, bundle, bulk
BO	Burn out
c	Condensation
cr	Critical
D	Drag
DNB	Departure from nucleate boiling
DO	Dry out
e	Computed at thermodynamic equilibrium
FDB	Fully developed boiling
f	Friction, fins
FC	Forced convection
g	Gas
go	Gas only
h	Heating, heated
i	Related to the ith component
in	Inlet, interface
iso	Isothermal
l	Liquid, laminar
lo	Liquid only
m	Mean
max	Maximum
min	Minimum
n	Nucleation
o	Outside, external, stagnant
PB	Pool boiling
pc	Pseudocritical
r	At radius r
stat	Stationary
sat	Saturation
sub	Subcooled
t	Turbulent
trans	Transient
tot	Total
tp, TP	Two-phase
v	Vapor
w	Wall, wetted
x, y, z	Related to x, y, z coordinates
∞	Ambient condition, remote

TABLE OF CONTENTS

Volume I

Volume II

Chapter 1

PHYSICAL PROPERTIES AND CHARACTERISTICS DIMENSIONLESS GROUPS

I. INTRODUCTION

Within the limits of this book, which is aimed at the most advanced aspects of heat transfer, only brief mention is made of the transport and of the physical properties of fluids. The properties of two-phase fluids are treated in greater detail; a list of dimensionless groups in use in thermohydraulic problems is given. For deeper insight, see the fundamental texts of physics.

II. UNITS OF MEASURE

All fluid properties, unless otherwise specified, are expressed using the International System of Units; in such a system, the four fundamental magnitudes, whose combination leads to the units of all transport and physical properties of interest in thermohydraulics, are length, L (m); mass, M (kg); time, t (sec); thermodynamic temperature, T (K; degrees Kelvin).

III. BASIC PROPERTIES

Fundamental properties, whose use is common in traditional single-phase flow problems, are here indicated as "basic properties". Properties typical of two-phase mixtures are considered in Section IV.

As a consequence of the well-known meaning of basic properties, only a list is reported, with the indication of the measure units (Table 1).

The nature of the basic properties has been classified as thermodynamic, physical (not thermodynamic), transport, and geometrical.

IV. TWO-PHASE MIXTURES' PROPERTIES

In the analysis of two-phase-mixtures physical and transport properties, only a particular situation is considered here, owing to its relevance in industrial applications: that of a one-component fluid, where a liquid phase and a gaseous (vapor) phase are simultaneously present.

In two-phase mixtures heat transfer problems, physical properties and transport parameters of the two single phases are not sufficient, the introduction of parameters characterizing the interaction between the phases being necessary.

Properties and parameters representing the average characteristics of the mixture, the mechanical, or thermal interactions between the two phases are considered here. Some of them are representative of a system dimension scale (global parameters), while others refer to a fluid-particles-dimension scale.

A. Void Fraction

Void fraction (volumetric gas fraction) is defined as the ratio between the flow area occupied by the gaseous phase and the overall flow area:

$$\alpha = A_g/A \tag{1}$$

Table 1
BASIC PROPERTIES (FOR A FLUID, ALL THERMODYNAMIC PROPERTIES ARE GENERAL FUNCTION OF TWO THERMODYNAMIC PROPERTIES)

Property	Symbol	Nature	Definition	Rel. with fund. magnitudes	Unit (SI)
Density (volumic mass)	ρ	Thermodynamic property	Mass of a substance contained in unit volume	M/L^3	kg/m^3
Specific weight	γ	Thermodynamic property	Weight/unit vol of a substance, $\gamma = \rho \cdot g$, g is gravity acceleration	$M/L^2\ t^2$	N/m^3
Specific heat	c; c_p	Thermodynamic property	Heat amt. necessary to increase temp of unit mass of a substance of 1 K, following a definite thermodynamic transformation (c_p may be referred to an isobaric transformation)	$L^2/t^2\ T$	J/(kg K)
Thermal conductivity	k	Thermodynamic property	Proportionality coefficient between local heat flux and temp gradient (only conduction); as a consequence, it represents thermal flux which causes (in a material) a unit spatial gradient of temp	$ML/t^3\ T$	W/(m k)
Thermal diffusivity	α	Thermodynamic property	$\alpha = k/(\rho\ c_p)$	L^2/t	
Viscosity (dynamic)	μ	Thermodynamic property	Proportionality coefficient between shear stress and local velocity gradient; as a consequence, it represents shear stress to apply to a fluid to obtain a unit velocity gradient	M/L t	$N.s/m^2$ Pa·s; ng/ms
Viscosity (Kinematic)	ν	Thermodynamic property	$\nu = \mu/\rho$	L^2/t	m^2/s

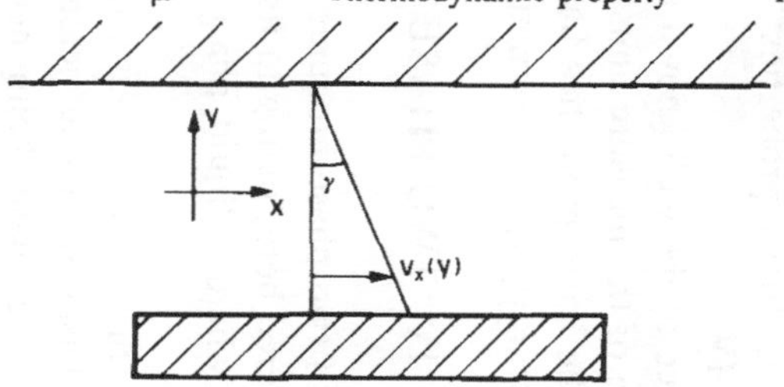

$$\tau_{xy} = \mu \frac{dv_x}{dy}$$

$$\tau_{xy} = \mu \frac{d\gamma}{dt}$$

Surface tension	σ	Thermodynamic property (only liquids)	Mechanical work necessary to increase free surface area of a liquid of one unit, through movements of molecules	M/t^2	N/m
Specific enthalpy	i	Thermodynamic property	$i = u + pv$	L^2/t^2	J/kg
Specific energy (internal energy)	u	Thermodynamic property	$du = \delta Q = \delta L$	L^2/t^2	J/kg
Specific entropy	s	Thermodynamic property	$ds = \delta Q/T$	L^2/t^2	J/kg
Pressure	p; P	Thermodynamic and transport property	$p = F/A$	$M/L\ t^2$	Pa; N/m^2; $kg/(m/sec^2)$
Velocity	v	Transport property	$v = dl/dt$	L/t	m/sec
Area	A	Geometrical property	Measure of a surface	L^2	m^2
Mass flow rate	Γ	Transport property	$\Gamma = \rho\ v\ A$	M/t	kg/sec
Volumetric flow rate	Q	Transport property	$Q = vA = \Gamma/\rho$	L^3/t	m^3/sec
Mass velocity (specific mass flow rate)	G	Transport property	$G = \rho v = \Gamma/A$	M/L^2t	$kg/(m^2/sec)$
Volume	V	Geometrical property	Space portion occupied by a substance	L^3	m^3
Specific volume (unit mass volume)	v	Thermodynamic property	$v = l/\rho$	L^3/M	m^3/kg
Elasticity (Young module)	E	Physical property	Parameter measuring volumetric elasticity of a substance: $E = -dp/(dV/V)$	M/Lt^2	Pa
Compressibility	l/E	Physical property	Parameter measuring relative variation of volume of a substance under effect of external forces	Lt^2/M	l/Pa
Diffusivity (mass)	D	Physical property	Proportionality coefficient between diffusion molar flux (of substance A) and *M* conc gradient $J_{n,A} = -CD_A\ \partial X_A/\partial n$ (n = normal; c = total *M* conc; X_A = substance A *M* fraction)	L^2/t	m^2/sec
Expansion	β	Physical property	$\beta = \frac{1}{v}\left(\frac{\partial v}{\partial T}\right)_p = -\frac{1}{\rho}\left(\frac{\partial \rho}{\partial T}\right)_p$	l/T	K^{-1}

Naturally, α may assume values between 0 (liquid only) and 1 (gas only). Such a definition refers to average values in time and in space, since locally and instantaneously, void fraction may differ considerably from the average value.

B. Vapor Quality

Vapor quality is defined as the ratio between the gaseous mass flow rate (averaged on the whole flow area) and the total mass flow rate (in static conditions, it is the ratio between vapor mass and total mass):

$$X = \Gamma_g/\Gamma \qquad (2)$$

It assumes numerical values between 0 and 1. The complement to 1 is sometimes indicated as "humidity quality":

$$1 - X = \Gamma_l/\Gamma \qquad (3)$$

C. Volumetric Vapor Quality

The volumetric vapor quality is the ratio between the gaseous-phase volumetric flow and the total volumetric flow:

$$\beta = Q_g/Q \qquad (4)$$

The relationship between volumetric vapor quality and vapor quality is

$$\beta = 1/\{1 + [(1 - X)/X] \cdot \rho_g/\rho_l\} \qquad (5)$$

D. Thermodynamic Quality

In one-component two-phase systems (liquid + vapor) thermodynamic quality is defined on the basis of an enthalpic balance (at equilibrium):

$$X_i = (\bar{i} - i_l)/(i_v - i_l) \qquad (6)$$

Where $\bar{i}$ is the mixture average enthalpy, $\bar{i}_l$ is the saturated-liquid-phase enthalpy at the same pressure of the mixture, and i_v is the enthalpy of the saturated vapor phase under the same pressure conditions. Only when the two phases of the same component are in thermodynamic equilibrium is the relation $X = X_i$ valid. Situations of thermodynamic disequilibrium (phases with different temperatures) may occur; a typical example is that of subcooled boiling, where a vapor phase is present in spite of a negative value of the thermodynamic quality; in such a situation, it is, obviously, $X > X_i$. Thermodynamic quality may assume negative values and values greater than 1.

E. Velocity

In two-phase systems, the parameter "velocity" is generally distinct for the two phases. The gaseous phase velocity is defined as:

$$v_g = Q_g/A_g = G_g/\rho_g \qquad (7)$$

where Q_g, G_g are, respectively, the volumetric flow and the mass velocity of the gaseous phase. A corresponding relation is valid for the liquid phase; v_g and v_l are defined with reference to an average value on the flow area. A mixture velocity is also defined as:

$$v_{mix} = \Gamma/(\rho_{mix} \cdot A)$$

F. Superficial Velocity

The superficial velocity is defined, for each of the two phases, as follows:

$$J_g = Q_g/A; \quad J_l = Q_l/A \tag{8}$$

It is a parameter averaged in time, not corresponding to a real velocity, and making reference to the whole flow area available. It is used in technical literature to characterize, from a hydraulic point of view, the two-phase flow. The relations between void fraction, "real" velocity, and superficial velocity are

$$J_g = \alpha v_g; \quad J_l = (1 - \alpha)\, v_l \tag{9}$$

A mixture superficial velocity is also used:

$$J = J_g + J_l = \alpha v_g + (1 - \alpha)\, v_l$$

G. Slip Ratio

The slip ratio between two phases is defined as:

$$S = v_g/v_l \tag{10}$$

If the average values of the two velocities on the transverse area are used, the slip ratio is considered averaged on the same area; its value may differ very much from local values.

The slip ratio is generally greater than 1, typically in the situation of cocurrent two-phase flows.

In water systems, S assumes values of some units (2 to 5) at high pressure and values of the order of 100 at low pressure, when the difference between the densities of the two phases is highest. Void fraction, vapor quality, and slip ratio are not independent.

In a one-dimensional treatment of the flow, when it is possible to make reference to average values on the flow area, the following relationships hold:

$$\alpha = 1/\{1 + [(1 - X)/X]\, S\, \rho_g/\rho_l\} \tag{11}$$

$$X = 1/\{1 + [(1 - \alpha)/\alpha]\,(1/S)\, \rho_l/\rho_g\} \tag{12}$$

$$S = [X/(1 - X)]\,[(1 - \alpha)/\alpha]\, \rho_l/\rho_g \tag{13}$$

H. Mass Flow Rate

The same definition is adopted as in single-phase flows:

$$\Gamma_g = \rho_g\, v_g\, A_g; \quad \Gamma_l = \rho_l\, v_l\, A_l \tag{14}$$

For the total mass flow rate, it is

$$\Gamma = \Gamma_g + \Gamma_l \tag{15}$$

I. Volumetric Flow Rate

The same definition is valid as in single-phase flows:

$$Q_g = v_g\, A_g = \Gamma_g/\rho_g; \quad Q_l = v_l\, A_l = \Gamma_l/\rho_l \tag{16}$$

The total volumetric flow rate is

$$Q = Q_g + Q_l \tag{17}$$

J. Mass Velocity (Specific Mass Flow Rate)

The same definition is valid as in single-phase flows:

$$G = \Gamma/A \tag{18}$$

$$G_g = \Gamma_g/A_g; \quad G_l = \Gamma_l/A_l \tag{19}$$

It is a very important parameter, widely used in two-phase flows with a change of phase. Total mass velocity is different from the sum of the mass velocities of the single phases; for this reason, the apparent mass velocities are also used.

K. Apparent Mass Velocity

This velocity is defined for each phase, with reference to the whole flow area:

$$G_g^* = \Gamma_g/A; \quad G_l^* = \Gamma_l/A \tag{20}$$

The following relationships are valid:

$$G = G_g^* + G_l^* \tag{21}$$

$$G_g^* = \alpha\, G_g = \alpha\, \rho_g\, v_g = \rho_g\, J_g \tag{22}$$

$$G_l^* = (1 - \alpha)\, G_l = (1 - \alpha)\, \rho_l\, v_l = \rho_l\, J_l \tag{23}$$

L. Density and Specific Volume

The density of a two-phase mixture may assume all the values between the gaseous-phase and the liquid-phase densities. Generally, for a one-component mixture, the difference between the extreme values increases if pressure decreases from its value at the critical point.

Density of two-phase mixtures may vary in the direction of the flow (as a consequence of pressure variation and heat transfer) and transversally to the direction of the flow (owing to the particular distribution of the two phases). In the most common applications, it is not necessary to know the local value of density, being sufficient the knowledge of an average value on the flow area.

The density of a two-phase flow may be defined as the ratio between the mass contained in a portion of space and its volume. Obviously, owing to the fluctuations typical of two-phase mixture flow regimes, reference is made to average values of density with respect to time.

The evaluation of a two-phase mixture density may be done using the parameter "void fraction", representing the fraction of the unit of volume occupied by the gaseous phase (See Section IV.A). The following relationship applies:

$$\bar{\rho} = \alpha\rho_g + (1 - \alpha)\, \rho_l \tag{24}$$

The average specific volume $\bar{v}$ is defined as

$$\bar{v} = 1/\bar{\rho}$$

In technical literature, another parameter is defined: the "flow density" ($\bar{\rho}_f$). It is the density of a hypothetical single-phase fluid that flowing with a volumetric flow rate equal to that of the two-phase flow, also has the same mass flow rate.

By the definition:

$$\bar{\rho}_f = \Gamma/Q = \rho_l\, Q_l/Q + \rho_g\, Q_g/Q \tag{25}$$

the following relationships are valid:

$$\bar{\rho}_g = (1 - \beta)\, \rho_l + \beta\rho_g \tag{26}$$

$$\bar{\rho}_f = 1/[X/\rho_g + (1 - X)/\rho_l] \tag{27}$$

$$\bar{v}_f = 1/\bar{\rho}_f = Xv_g + (1 - X)\, v_l \tag{28}$$

The two parameters, $\bar{\rho}$ and $\bar{\rho}_f$, are both widely used. The average density $\bar{\rho}$ is generally used in energy balances while the flow density, $\bar{\rho}_f$, is used in momentum balances.

The use of the flow density, $\bar{\rho}_f$, is easier (the validity of its application must be verified) because it depends only on the vapor quality that may be deduced through simple thermal balances. On the contrary, the average density $\bar{\rho}$ requires the knowledge of the void fraction. This parameter cannot be evaluated immediately because, even knowing the values of the densities for the two phases and the quality, the necessity of knowing the actual value of the slip ratio remains. Three typical situations, depending on the values assumed by the slip ratio, may occur:

$$\text{a) } S = 1; \quad \alpha = \beta; \quad \bar{\rho} = \bar{\rho}_f \tag{29}$$

Such a situation is typical of homogeneous mixtures.

$$\text{b) } S < 1; \quad \alpha > \beta; \quad \bar{\rho} < \bar{\rho}_f \tag{30}$$

The gaseous phase has a lower velocity than the liquid phase; such a situation may occur in a vertical downflow or in the presence of subcooled boiling, where the gaseous phase is concentrated in the peripheric area of the duct, with a velocity lower than the average velocity.

$$\text{c) } S > 1; \quad \alpha < \beta; \quad \bar{\rho} > \bar{\rho}_f \tag{31}$$

It is the most common situation.

M. Viscosity

In two-phase mixtures' thermohydraulic calculations, the necessity of evaluating a characteristic viscosity may occur. A typical example is the evaluation of friction pressure drops using the homogeneous model, where the use of a mixture Reynolds number requires the definition of a mixture viscosity. Various authors have proposed different correlations for the evaluations of such a parameter:

- Owens:

$$\mu_{TPF} = \mu_l \tag{32a}$$

- Dukler:

$$\mu_{TPF} = [\mu_l(1 - X)\, v_l + \mu_g\, X\, v_g]/[v_l(1 - X) + v_g\, X] \tag{32b}$$

- Lombardi and Silvestri:

$$\mu_{TPF} = X\mu_g + (1 - X)\,\mu_l \tag{32c}$$

- Davidson:

$$\mu_{TPF} = \mu_l[1 + X(v_g/v_l - 1)] \tag{32d}$$

- Isbin:

$$1/\mu_{TPF} = X/\mu_g + (1 - X)/\mu_l \tag{32e}$$

V. DIMENSIONLESS GROUPS

The use of dimensionless groups (or numbers) is based on the verification, for specific phenomena of interest in thermo-fluidodynamics, of similarity situations. In such situations, different fluids in different geometrical and physical conditions, but with the same values assumed by particular groups of parameters (physical and geometrical), show analogous behavior. A list of the most used dimensionless groups in thermohydraulic problems is reported, with an indication of their meaning and field of application, grouped in two gross wholes: fluidodynamics and heat transfer. Some dimensionless groups such as the Reynolds number, are of interest both in fluidodynamic and in heat transfer problems.

In order to facilitate the comprehension of the physical meaning of the various dimensionless groups, the connections existing among the fundamental physical parameters and the intervening forces, whose role is fundamental in the definition itself of the dimensionless groups are recalled. All the forces are expressed in N (kg · m/sec^2).

- Inertial force:

$$F_i \propto \rho\, v^2\, A \tag{33}$$

- Elastic force:

$$F_e \propto \mu\, v^2\, t \tag{34}$$

- Viscous force:

$$F_v \propto \mu\, v\, A/L \tag{35}$$

- Buoyancy force:

$$F_b \propto \rho\, \beta\, L^3\, g\, \Delta T \tag{36}$$

- Gravity force:

$$F_g \propto \rho\, g\, AL; \quad \propto \Delta\rho\, g\, A\, L \tag{37}$$

- Centrifugal force:

$$F_c \propto \rho\ v^2\ L^3/R \quad (38)$$

- Surface tension force:

$$F_S \propto \sigma\ L \quad (39)$$

- Compression force:

$$F_{CO} \propto E\ L^2 \quad (40)$$

- Shear force on the wall:

$$F_A \propto \tau_w\ L^2 \quad (41)$$

A. Fluidodynamics

1. Friction Factor

The force exerted by a fluid moving in a duct on the containment surface may be subdivided in two terms: a force, F_o, that the fluid would exert even if it were in rest, and a force, F_A, associated to the motion of the fluid. This last term is oriented in the same direction of the average velocity $\bar{v}$ in the duct.

A dimensionless group known as "friction factor" (f_f), is defined as the ratio between the force, F_A, and the inertial force F_i:

$$f_f = F_A/F_i \quad (42)$$

Practically speaking, the friction factor represents the friction pressure losses in a duct (with a diameter d and a length L ($\propto$), F_A) $\propto$ expressed in terms of kinetic heights (α F_i):

$$f_f = [\Delta p/(\rho\ \bar{v}^2/2)]\ d/L \quad (43)$$

In the literature, another friction factor is defined, which is widely used in applications (Chapter 4), called the "Fanning" friction factor (f), related to f_f by the relationship:

$$f = f_f/4 \quad (44)$$

2. Drag Coefficient

Drag coefficient, C_d, is defined as the ratio between gravity and inertial forces for a body moving within a fluid:

$$C_d = F_g/F_i \propto \Delta\ \rho\ g\ A\ L/(\rho\ v^2\ A) = \Delta\ \rho\ g\ L/(\rho\ v^2) \quad (45)$$

where ρ is the fluid density, $\Delta\rho$ is the difference of density between the fluid and the body, v is the relative velocity, and L is a geometrical dimension, characteristic of the body itself.

3. Bingham Number

The Bingham number characterizes the behavior of some non-newtonian fluids (Chapter 16) and is defined as the ratio between the internal shear force and the viscous force:

$$B_m = \tau_y\ L/(\mu_o\ v) \quad (46)$$

where μ_o is the rigidity coefficient of the fluid, having the same dimensions of a viscosity.

4. Blake Number

The Blake number has the same physical meaning as the Reynolds number, but in the situation of solid particles beds. It is

$$B = F_i/F_v = v\ \rho/[\mu(1 - e)\ S] \tag{47}$$

where S is the ratio between the area of the particles' surface and their volume (L^2/L^3) and e is the void fraction of the bed.

5. Bond Number

The Bond number is of interest in atomization processes and in situations of drag of drops or of solid particles. It may be considered as the ratio between the gravity force and the surface tension force:

$$Bo = F_g/F_s \propto \Delta\ \rho\ g\ L^3/(\sigma\ L) = (\rho_p - \rho)\ L^2\ g/\sigma \tag{48}$$

where ρ_p is the particles' density and ρ is the fluid density.

6. Capillarity Number

The capillarity number is used in problems of atomization and of two-phase flows in fluid beds. It represents the ratio between the viscous force and the surface tension force:

$$Ca = F_v/F_s \propto (\mu\ v\ A/L)/(\sigma\ A/L) = \mu\ v/\sigma \tag{49}$$

7. Cauchy Number

It intervenes in problems with compressible fluids. It is the ratio between inertial and compressibility forces:

$$C = F_i/F_{co} \propto \rho\ v^2\ L^2/(EL^2) = \rho\ v^2/E \tag{50}$$

8. Dean Number

The Dean number characterizes the flow of a fluid in a curved duct. It is representative of the ratio between centrifugal and inertial forces. It depends, moreover, on the fluido-dynamic characteristics of the flow through the Reynolds number:

$$D = (\rho\ v\ L/\mu)\ (L/2R)^{0.5} \tag{51}$$

where R is the curvature radius in the duct.

9. Euler Number

The Euler number is used in applications regarding pressure losses due to friction in ducts. It is defined as:

$$Eu = \Delta\ p/(\rho\ v^2) \tag{52}$$

and is the ratio between the friction pressure loss and the kinetic height characterizing the fluid motion.

10. Elasticity Number

The elasticity number characterizes viscoelastic fluid flows (see Chapter 16) and is defined as the ratio between the elastic and inertial forces:

$$El = \mu\ t_r/(\rho\ L^2) = (\mu^2\ v^2/G_t)/(\rho\ v^2\ L^2) \quad (53)$$

t_r being the fluid relaxation time and G_t is the transveral elasticity module.

11. Froude Number

The Froude number is important in fluidodynamic problems where a free surface is present or, generally, where gravitational forces become important (superficial waves). It is an index of the ratio between inertial and gravity forces:

$$Fr = F_i/F_g \propto v^2/(g\ L) \quad (54)$$

12. Galileo and Archimede Numbers

The Galileo number is used in the study of thermohydraulic phenomena where gravitational forces become important. It is defined as:

$$Ga = F_i\ F_g/(F_v)^2 \propto \rho^2\ g\ L^3/\mu^2 \quad (55)$$

It is connected to the Reynolds and Froude numbers:

$$Ga = Re^2/Fr \quad (56)$$

In the presence of bubbles or in the study of their formation, a different form of the Galileo number is sometimes used, called the Archimede number:

$$Ar = [\rho_l\ g\ L^3(\rho_l - \rho_g)]/\mu^2 \quad (57)$$

In such a form, the bubble-free rise velocity is connected to the Archimede number. The forces ratio is the same as in the Galileo number, where the gravitational force is substituted by the Archimede force:

$$F_g = F_{Ar} = g(\rho_l - \rho_g)\ AL \quad (58)$$

The Archimede number may also characterize the constitution and the development of gravitational waves in a two-phase viscous fluid: as a consequence of the substitution in the Galileo number of the L dimension with the Laplace length, $L = \sqrt{\sigma/[g(\rho_l - \rho_g)]}$, it is

$$Ar = (\rho_l^{1/2}\ \sigma^{3/2})/\{\mu^2[g(1 - \rho_g/\rho_l)^3]^{1/2}\} \quad (59)$$

13. Knudsen Number

It is an index of the rarefaction of a gas and is defined as the ratio between the molecules' average free path and a characteristic length of the gas flow:

$$Kn = x_m/L \quad (60)$$

If $Kn << 1$, the flow is continuous, while for $Kn >> 1$, a free molecular current occurs because the gas transport phenomena are practically independent of the molecular interaction.

14. Kutateladze Number

The Kutateladze number has a relevance in fluidodynamic problems where two-phase mixtures are involved. It is defined as:

$$Ku = G\ X/\{\rho_g^{1/2}[g\ \sigma(\rho_l - \rho_g)]^{1/4}\} \quad (61)$$

It depends on the ratio between the inertial and the superficial tension forces:

$$(Ku)^2 = F_i/F_s \propto \rho\ v^2\ L^2/(\sigma\ L) \tag{62}$$

Ku expression may be obtained if in Equation 62 the following substitutions are made:

$$\rho \rightarrow \rho_g; \quad v \rightarrow J_g; \quad L = \sqrt{\sigma/[g(\rho_l - \rho_g)]} \text{ (Laplace length)}$$

The Kutateladze number is used, for example, as a criterion for the occurrence of flooding phenomena (See Chapter 3).

15. Mach Number

The Mach number is generally used in the study of the motion of a compressible fluid. It is an index of the compressibility of the fluid and is given by the ratio between a system characteristic velocity and the sound velocity in the medium:

$$M = v/a \tag{63}$$

It is representative of the ratio between the inertial and the elastic forces, being:

$$(M)^2 = F_i/F_{co} \propto \rho\ v^2\ A/(EA) = \rho\ v^2/E \tag{64}$$

Following the Laplace equation, the sound velocity may be expressed as:

$$a = \sqrt{E/\rho} \tag{65}$$

from which Equation 63 may be obtained.

Since it is

$$\Delta\rho/\rho_o = \Delta\ p/E \approx \rho_o\ v^2/(2E) = M^2/2$$

a fluid may be considered incompressible if $\Delta\rho/\rho_o << 1$, where $\Delta\rho$ is the variation of density due to pressure variation Δp.

16. Martinelli Number

The Martinelli number is used in the fluidodynamic study of two-phase flows and is related to the friction pressure gradients relative to the flows of the liquid and the vapor phases, independently considered:

$$X_{tt} = [(dp/dz)_l/(dp/dz)_g]^{1/2} = (\rho_g/\rho_l)^{0.5}\ (\mu_l/\mu_g)^{0.1}\ [(1 - X)/X]^{0.9} \tag{66}$$

Other two-phase flow parameters, as the void fraction, may be related to the Martinelli parameter.

17. Ohnesorge Number

The Ohnesorge number relates the viscous, inertial, and the surface tension forces in problems of dragged particles fluidodynamics:

$$Z = F_v/(F_i\ F_s)^{0.5} \propto \mu/(\rho\ \sigma\ L)^{0.5} \tag{67}$$

The following relationship among the Ohnesorge, Weber, and Reynolds numbers applies:

$$Z = (We)^{0.5}/Re \tag{68}$$

18. Reynolds Number

The Reynolds number is certainly the most relevant and used dimensionless group in applications. It characterizes the motion of a fluid from a hydrodynamic, geometrical, and physical point of view. It is very important both in fluidodynamic correlations (friction factor) and in thermal correlations (heat transfer coefficient). It may be considered to be the ratio between the inertial and the viscous forces acting on the fluid:

$$Re = F_i/F_v \propto \rho V^2 A/(\mu V A/L) = \rho V L \mu \tag{69}$$

where L is the geometrical parameter characteristic of the flow.

Typically, Reynolds numbers lower than 2100 characterize laminar flow regimes, where viscous forces prevail over inertial forces, while Reynolds numbers greater than 5000 indicate turbulent flows, where inertial forces are predominant. Intermediate values (transition zone) are indicative of a generally unstable flow.

19. Schmidt Number

The Schmidt number is introduced in mass transport problems and has a meaning analogous to that of the Prandtl number, in the case of heat transfer problems.

It represents the ratio between the momentum diffusivity which affects momentum transmission as an effect of viscosity when a velocity gradient occurs, and the mass diffusivity between two different chemical substances, which affects the mass flux when a concentration gradient occurs:

$$Sc = \mu/(\rho D_{AB}) = \nu/D_{AB} \tag{70}$$

20. Sherwood Number

In processes involving mass transfer, the Sherwood number is analogous to the Nusselt number introduced in heat transmission. The Sherwood number is indicative of the ratio between the entity of mass transfer in a binary mixture through an interface and the mass transport due to molecular diffusion:

$$Sh = k_x \cdot d/(CD_{AB}) \tag{71}$$

k_x being the mass transfer coefficient, C, the molecular concentration, and D_{AB}, the binary diffusion coefficient.

21. Strouhal Number

The Strouhal number intervenes in periodic oscillatory phenomena characteristic of a turbulent flux, in the interaction with a submerged body. It is defined as the reciprocal of the distance between two subsequent vortices, expressed in terms of the obstacle characteristic dimension (f: frequency of oscillations):

$$Sl = fL/V \tag{72}$$

It is also used to characterize a vibrating structure under the forcing action of a turbulent flow, where f indicates the structure's natural vibration frequency.

22. Weber Number

It is an adimensional group intervening in problems concerning the interaction between two different phases, as in the bubble's formation or in the rupture in droplets of a liquid jet on the free surface.

It is representative of the ratio between the inertial and the surface tension forces:

$$We = F_i/F_s \propto \rho V^2 L^2/(\sigma L) = \rho V^2 L/\sigma \quad (73)$$

23. Prandtl Velocity Ratio

In the analysis of the universal velocity profile an adimensional velocity is used, v^+, which represents the square root of the ratio between the inertial and the friction forces on the wall of a duct, with a fluid in turbulent flow:

$$(v^+)^2 = F_i/F_A \propto \rho V^2 A/(\tau_W A); \quad v^+ = v/(\tau_W/\rho)^{1/2} \quad (74)$$

B. Heat Transfer

1. Biot Number

The Biot number is used in problems regarding the sudden contact of a fluid with a wall at a quite different temperature (thermal shock). It is defined as:

$$Bi = h\,L/k \quad (75)$$

where L is a system characteristic length, for example, the thickness of a slab subject to the shock. The physical interpretation of the Biot number is analogous to that of the Nusselt number; it is representative of the relationship between the transport of heat between the fluid and the wall and the transport of heat inside the wall, due to the wall material thermal conduction (in the Nusselt number, k is the fluid thermal conductivity, while the Biot number is the wall's thermal conductivity).

2. Brinkman Number

The Brinkman Number is defined as:

$$Br = \mu\, v^2/(g\, k\, \Delta T) \quad (76)$$

It is a measure of the importance of heating (in a moving fluid) due to viscosity, with respect to heat transfer resulting from the application of a difference of temperature.

Generally, the viscous heating is not relevant; it becomes important when relevant velocity variations occur in a very reduced space, as in:

1. Motion of lubricant between solid parts in a very fast relative motion
2. Motion of plastic substances through draw plates in high-velocity extrusions
3. Motion of air in proximity of the surface of a satellite or of a rocket (reentering the atmosphere)

A small value of the Brinkman number indicates that the heat produced by viscous dissipation may be transferred (or removed) by thermal conduction.

3. Condensation Number

The condensation number is defined as:

$$Co = (h/k)\,[\mu^2/(\rho^2 g)]^{1/3} \quad (77)$$

and is used in the study of similarity criteria in condensation processes where the interaction among capillarity, gravity, and viscosity is important. It may be obtained from the following relationship:

$$\mathrm{Co} = (q''_{eff}/q''_k)\ [(F_v)^2/(F_i\ F_g)]^{1/3} \tag{78}$$

where q''_{eff} is the effective thermal flux and q''_k is the thermal flux due to conduction:

$$q''_{eff} \propto h\ \Delta T; \quad q''_k \propto (k/L)\ \Delta T$$

Introducing the Nusselt and the Galileo numbers it is possible to write:

$$\mathrm{Co} = \mathrm{Nu}/(\mathrm{Ga})^{1/3} \tag{79}$$

4. Eckert Number

The Eckert number is defined as:

$$\mathrm{Ec} = v^2/(g\ c_p\ \Delta T) \tag{80}$$

This adimensional group may be interpreted differently if the fluid is incompressible or not. If the fluid is incompressible, the Eckert number may be interpreted as the ratio between the Brinkman and Prandtl numbers:

$$\mathrm{Ec} = \mathrm{Br}/\mathrm{Pr} \propto [\mu\ v^2/(g\ k\ \Delta T)] \cdot [k/(c_p\ \mu)] \tag{81}$$

In this situation, as for the Brinkman number, the Eckert number is an index of the degree of viscous dissipation. The production of heat due to friction becomes significant when the Eckert number is nearly 1.

If the fluid is compressibile, the Eckert number is representative of the importance of heat produced by friction and compression in the definition of the temperature profile in the fluid near the wall.

5. Fourier Number

The Fourier number allows a comparison between heat transferred by internal conduction and heat absorbed by the body. It is defined as:

$$\mathrm{Fo} = k\ A\ \Delta T\ t/(\rho\ c_p\ AL^2\ \Delta T) = kt/(\rho\ c_p\ L^2) \tag{82}$$

6. Graetz Number

In a laminar flow with constant properties of the fluid, heat transfer correlations provide the Nusselt number as a function of a characteristic product:

$$\mathrm{Gz} = (\pi/4)\ \mathrm{Re}\ \mathrm{Pr}(d/L) \tag{83}$$

which is called the Graetz number. It is generally expressed as a function of the mass flow rate:

$$\mathrm{Gz} = \Gamma\ c_p/(kL) \tag{84}$$

Its physical meaning may be reconducted to the ratio between convection and conduction heat transfer.

7. Grashof Number

The Grashof number characterizes heat transfer in natural convection. It may be physically interpreted as the ratio between a gravitational force (buoyancy) due to density differences (caused by temperature gradients) and a viscous friction force:

$$Gr = [F_b]/[F_v] \propto \beta \, \Delta T \, g \, \rho^2 \, L^3/\mu^2 \tag{85}$$

8. Jacob Number

The Jacob number is characteristic of phase change problems and represents a thermal fluidodynamic similarity criterion in phase change situations. It is defined as:

$$Ja = [c_{p,l}(T_w - T_{sat})/\lambda] \, (\rho_l/\rho_v) \tag{86}$$

representing a sort of efficiency of phase change that occurs, for example, in proximity to a solid wall at temperature T_w for a fluid at saturation temperature T_{sat} with latent heat λ.

9. Lewis Number

The Lewis number is the ratio between the mass diffusivity and the thermal diffusivity:

$$Le = Pr/Sc = D/\alpha \tag{87}$$

and is representative of the ratio between mass and heat transfer at the separation interface between a liquid and a gas or in a two-component gas mixture for which, generally, Le is not far from 1.

10. Marangoni Number

On a free or interface surface in nonisothermal conditions, a surface tension gradient $\partial\sigma/\partial T$ occurs. The surface tension depends on the viscous forces in the superficial layer and is determined by the temperature distribution, i.e., by a characteristic temperature difference ΔT, by the conductivity coefficient k, and by the heat capacity ρc_p. The dimensionless group that derives from such parameters is the Marangoni number:

$$Ma = [(\Delta T \cdot L)/(\mu \, \alpha)] \cdot (\partial \, \sigma/\partial T) \tag{88}$$

11. Nusselt Number

The Nusselt number may be interpreted as the ratio between convection heat transfer (proportional to $h\Delta t$) and heat transfer due to internal conduction ($\propto$ (k/L) ΔT):

$$Nu = (h \cdot \Delta T)/[(k/L) \, \Delta T] = hL/k \tag{89}$$

and is widely used in heat transfer correlations for the evaluation of the heat transfer coefficient. The following relationship has a general validity: Nu = f (Re, Pr, Gr, Br), where Re, Reynolds number; Pr, Prandtl number; Gr, Grashof number; and Br, Brinkman number. In most practical situations, the effects of viscous dissipation are very small so that the Brinkman number may be neglected.

12. Péclet Number

The Péclet number is defined as the product of the Reynolds and Prandtl numbers:

$$Pe = Re \cdot Pr = \rho \, v \, L \, c_p/k \tag{90}$$

It assumes a great importance in liquid metals' heat transfer correlations. Liquid metals have a very low Péclet number owing to their high thermal conductivity.

To indicate the physical meaning of the Péclet number, it may be written:

$$Pe = (\rho v\, c_p\, \Delta T)/[k(\Delta T/L)] \qquad (91)$$

where Pe is the ratio between convection and conduction heat transfer. The Péclet number allows a classification of heat transfer modes:

1. If $Pe < 100$, heat transmission occurs exclusively through conduction
2. If $Pe \simeq 1000$, conduction transmission is comparable with convection transmission
3. If $Pe > 50000$, heat transfer occurs only through convection

13. Prandtl Number

The Prandtl number relates heat transmission with the fluid motion. It represents the ratio between the momentum diffusivity (kinematic viscosity) that regulates the transmission of momentum as an effect of viscosity when there is a velocity gradient, and the thermal diffusivity ($\alpha = k/(c_p\rho)$ that regulates heat transmission due to internal conduction when a temperature gradient is present:

$$Pr = \nu/[k/(c_p\, \rho)] = [\mu/\rho]/[k/(c_p\, \rho)] \qquad (92)$$

The Prandtl number characterizes completely the fluid nature (it may be considered a thermodynamic property) and does not include any parameter relative to a particular thermal fluidodynamic situation or to a particular geometrical condition.

14. Rayleigh Number

The Rayleigh number characterizes natural convection heat transfer. It is given by the product of the Prandtl and Grashof numbers:

$$Ra = Pr\, Gr = (c_p\, \mu/k)\, (\beta g\, \Delta T\, \rho^2\, L^3/\mu^2) \qquad (93)$$

It is indicative of the ratio between the gravity force, which affects fluid motion and heat transfer in natural convection problems, and the thermal molecular diffusivity.

The Rayleigh number also allows the establishment of the beginning of a convective motion, the applicability limit of the boundary-layer theory, and the transition between laminar and turbulent flow.

15. Richardson Number

The Richardson number is also called the stratification parameter because it is relative to stability problems in nonhomogeneous fluid flows (stratification). The influence of density variations on the stability of a fluid flowing on a horizontal slab is considerable in the situation of fluid heating from the bottom, i.e., when the density decreases in the direction of the gravitational force. This phenomenon is very important in metheorological processes. The Richardson number is defined as:

$$Ri = [(-g/\rho)\, (\partial \rho/\partial y)]/(\partial v/\partial y)^2_w \qquad (94)$$

The positive direction of y coordinate is upward. A small value of the Richardson number characterizes a homogeneous flow. If $Ri > 0$, the flow is stable, while if $Ri < 0$, it is unstable.

16. Stanton Number

The Stanton number represents the ratio between the amount of heat effectively exchanged and the maximum amount theoretically exchangeable in a heat transfer process with a flowing fluid (heat transfer efficiency). It is defined as:

$$St = Nu/(Re \cdot Pr) = h\Delta T/(\rho\ v\ c_p\ \Delta T) = h/(\rho\ v\ c_p) \tag{95}$$

where ΔT is the difference of temperature between the wall (T_w = uniform and constant) and the inlet fluid bulk.

17. Gr/Re² Ratio

The ratio between the Grashof number and the square value of the Reynolds number indicates the importance of the buoyancy forces with respect to the inertial forces and represents an index of the importance of natural convection:

$$\frac{Gr}{Re^2} = \frac{F_b}{F_i} \propto \frac{\rho\ \beta\ L^3\ g\ \Delta T}{\rho\ v^2\ L^2} = \frac{\beta\ \Delta T\ g\ L}{v^2} \tag{96}$$

REFERENCES

1. **Bird, R. B., Stewart, W. E., and Lightfoot, E. N.,** *Transport Phenomena*, John Wiley & Sons, New York, 1960.
2. **Perry, R. H. and Chilton, C. H.,** *Chemical Engineers' Handbook*, 5th ed., McGraw Hill, New York, 1973.
3. **Schlichting, H.,** *Boundary Layer Theory*, 7th ed., McGraw Hill, New York, 1979.
4. **Eckert, E. R. G. and Drake, R. M.,** *Analysis of Heat and Mass Transfer*, McGraw Hill, New York, 1972.

Chapter 2

FUNDAMENTAL CONSERVATION EQUATIONS IN FLUID SYSTEMS

I. INTRODUCTION

In fluid systems engineering, the solution of a great number of heat transfer and flow problems is obtained using empirical relationships whose validity, even justified by calculation rapidity needs, is generally limited to the specific flow configurations and restricted ranges for the main process parameters, corresponding to the situations in which they have been experimentally obtained. Furthermore, the experimental data scattering leads to the need of confidence intervals; if the phenomena are relevant to safety, empirical safety factors are generally necessary, to which no physical meaning may be attributed.

When a ''new'' situation must be analyzed, or a deep comprehension of the phenomenology is required, the solution of the thermal-fluidodynamic problem demands the application of the general conservation equations (of mass, momentum, and energy). Some semiempirical correlations are themselves special forms of the conservation equations, finalized to specific phenomenological situations.

The analytical solution of the conservation equations in their general form is not possible. The mathematical complexity affecting even simplified forms of them has limited, in the past, the use of general analysis methods based on such equations, in favor of empirical methods characterized by a narrow range of applicability.

The present wide diffusion of powerful numerical computers, together with the development of numerical solution techniques, makes the approaches based on conservation equations more popular. Presently in some specific applications, the conservation equations' techniques tend to prevail over even the empirical approaches, as in the analyses of complex two-phase phenomena (flooding, two-phase thermal-disequilibrium flows, etc.). In such situations, the use of conservation equations does not eliminate the need for specific testing in order to evaluate or quantify particular physical phenomena (i.e., energy interactions between the phases); it allows, nevertheless, the ability to neglect the contingent aspects generally connected to the specific test geometries. In the analysis of conservation equations, a distinction is made among single-phase laminar flows, single-phase turbulent flows, and two-phase flows.

Particular note must be made to momentum and energy equations: their exact solution requires a precise knowledge of the velocity profiles inside the flow channel. This limits an exact solution to single-phase, stabilized, laminar flows. In other situations of great interest in applications, such as turbulent flows, the chaotic nature of the flow does not allow an exact definition of the local velocities in the channel; even in the hypothesis of a numerical solution, extremely high detail in meshing ($>10^{12}$ points) would be required. That is why special approximate techniques have been developed, which, basing themselves upon statistical approaches, make the momentum and energy equations' solution possible, with an acceptable degree of approximation. Research is under way to ameliorate such techniques, in order that the validity of the proposed models may be extended.

The situation in two-phase flows is even more complicated, owing to the impossibility of defining the local velocities; many different approximate models for momentum (and energy) equations have been proposed which differ both in the simplifications adopted to simulate interphase interactions and especially in the choice of the parameters considered the most significant for expressing the flow evolution.

The status of knowledge is clearly lower in the case of two-phase phenomena with respect to single-phase phenomena. Further, important developments are expected in the field of

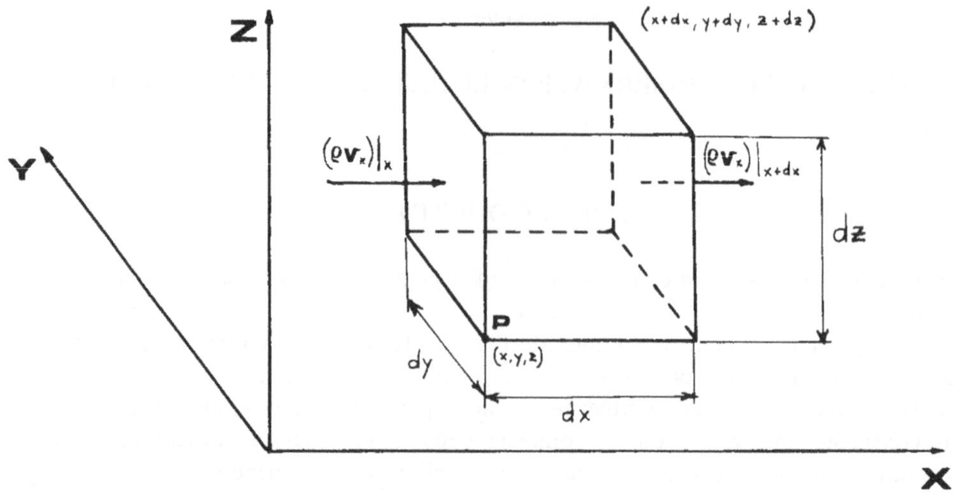

FIGURE 1. Reference control volume; continuity equation.

the representation and simulation of two-phase thermal-hydraulic phenomena. The models proposed up to now for two-phase flows are generally directly finalized at the specific application for which they have been developed (critical flows, liquid film interacting with a vapor flow, dispersed phase transport, etc.). For problems concerning single-phase turbulent flows, the choice regards the specific turbulence model to employ; in two-phase flows problems, the whole set of equations depends on the specific situation.

II. SINGLE-PHASE LAMINAR FLOW

In the majority of industrial applications, laminar flow conditions are not present; they are encountered only in quite special applications (non-newtonian fluids, very viscous oils, liquid metals). Nevertheless, these flow conditions are of relevant interest, allowing, in not-particularly complex geometries, the exact solution of the velocity, pressure, and temperature fields in the fluid system. It allows a rigorous solution of conservation equations and the extrapolation of a considerable amount of qualitative considerations to turbulent flows.

A. Continuity Equation

The continuity equation (mass conservation equation) expresses the instantaneous balance among mass entering the system (control volume), mass leaving the system, and mass remaining in the system (Figure 1)[1]:

(mass accumulation) = + (entering mass) − (leaving mass)

In punctual form, it may be written as:

$$\frac{\partial \rho}{\partial t} + \frac{\partial(\rho v_x)}{\partial x} + \frac{\partial(\rho v_y)}{\partial y} + \frac{\partial(\rho v_z)}{\partial z} = 0 \qquad (1)$$

It may also be written as:

$$\frac{\partial \rho}{\partial t} + v_x \frac{\partial \rho}{\partial x} + v_y \frac{\partial \rho}{\partial y} + v_z \frac{\partial \rho}{\partial z} + \rho\left(\frac{\partial v_x}{\partial x} + \frac{\partial v_y}{\partial y} + \frac{\partial v_z}{\partial z}\right) = 0 \qquad (2)$$

and, using the substantial derivative definition and that of the divergence operator, it may be written as (see Appendix):

$$\frac{D\rho}{Dt} + \rho \, \mathrm{div} \, \vec{v} = 0 \tag{3}$$

Two particular cases of applicative interest are that of incompressible fluids, where it may be written:

$$\mathrm{div} \, \vec{v} = 0 \tag{4}$$

and that of compressible stationary flows, where it may be written:

$$\mathrm{div}(\rho \vec{v}) = 0 \tag{5}$$

B. Momentum Equations

The momentum equations (flow equations) are obtained as an extension of the second law of dynamics, with the consideration that, in flowing fluids, the following forces may be present: (1) superficial forces (i.e., pressure), (2) volume forces (i.e., gravity), and (3) viscous forces (molecular diffusion and interaction).

In everyday applications, volume forces (and sometimes pressure superficial forces) are known as imposed "external" forces, while the viscous forces are dependent on the internal state of strain and therefore, on the fluid velocity field.

The most complex aspect of fluid dynamics is just the evaluation of the internal stress field, which expresses the mode of reaction of the fluid to imposed external stresses, causing deformation (strain field). As for solids, the relationship between the stress and the strain fields may be defined only by means of experimental tests and characterizes the fluid behavior. For newtonian fluids, the relationship between the stress and the strain fields was first defined by Stokes. Corresponding to the relation (identified by Hooke) for solids, of proportionality between the stress and the strain fields (hypothesis of elastic behavior valid in limited ranges of parameters), Stokes verified that for fluids, the stresses are proportional to the velocity gradients (shear rate) inside the fluid.

As a consequence, in analogy with the Hooke law, the following relationship may be introduced:

$$\begin{pmatrix} \tau_{xx} & \tau_{xy} & \tau_{xz} \\ \tau_{xy} & \tau_{yy} & \tau_{yz} \\ \tau_{xz} & \tau_{yz} & \tau_{zz} \end{pmatrix} = \mu \begin{pmatrix} \frac{\partial v_x}{\partial x} & \frac{\partial v_x}{\partial y} & \frac{\partial v_x}{\partial z} \\ \frac{\partial v_y}{\partial x} & \frac{\partial v_y}{\partial y} & \frac{\partial v_y}{\partial z} \\ \frac{\partial v_z}{\partial x} & \frac{\partial v_z}{\partial y} & \frac{\partial v_z}{\partial z} \end{pmatrix} +$$

$$+ \mu \begin{pmatrix} \frac{\partial v_x}{\partial x} & \frac{\partial v_y}{\partial x} & \frac{\partial v_z}{\partial x} \\ \frac{\partial v_x}{\partial y} & \frac{\partial v_y}{\partial y} & \frac{\partial v_z}{\partial y} \\ \frac{\partial v_x}{\partial z} & \frac{\partial v_y}{\partial z} & \frac{\partial v_z}{\partial z} \end{pmatrix} - \frac{2}{3} \mu \begin{pmatrix} \mathrm{div} \, \vec{v} & 0 & 0 \\ 0 & \mathrm{div} \, \vec{v} & 0 \\ 0 & 0 & \mathrm{div} \, \vec{v} \end{pmatrix} \tag{6}$$

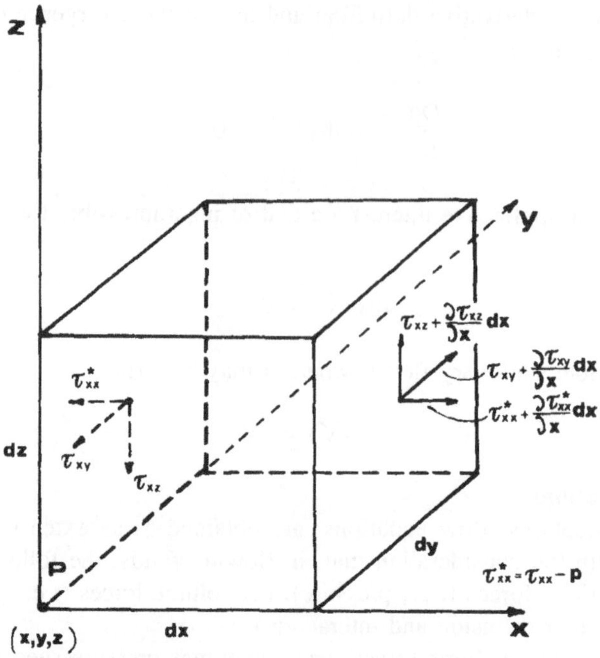

FIGURE 2. Stress field in a liquid volume. (Conventionally, a stress which the fluid on the side of an increasing value of the coordinate exerts on the fluid on the side with decreasing value of the coordinate, is counted positive when it acts in the direction in which the coordinate is counted positive. From this convention, the stresses indicated by the arrows in the figure are established as positive).

from which

$$\begin{cases} \tau_{x_i x_i} = \mu\left(2 \dfrac{\partial v_{x_i}}{\partial x_i} - \dfrac{2}{3} \operatorname{div} \vec{v}\right) \\ \tau_{x_i x_j} = \mu\left(\dfrac{\partial v_{x_i}}{\partial x_j} + \dfrac{\partial v_{x_j}}{\partial x_i}\right) \quad [i \neq j] \end{cases} \tag{7}$$

where: $i,j = 1, 2, 3$; $x_1 = x$, $x_2 = y$, $x_3 = z$.

Using the Stokes model, the general balance of momentum that may be expressed by (see Figure 2): (momentum accumulation) = + (entering momentum) − (leaving momentum) + (sum of forces acting on the system), assumes the following general form (with reference to x direction):

$$\rho \frac{Dv_x}{Dt} = -\frac{\partial p}{\partial x} + \frac{\partial}{\partial x}\left[2\mu \frac{\partial v_x}{\partial x} - \frac{2}{3} \mu \operatorname{div} \vec{v}\right] + \frac{\partial}{\partial y}\left[\mu\left(\frac{\partial v_x}{\partial y} + \frac{\partial v_y}{\partial x}\right)\right] + \frac{\partial}{\partial z}\left[\mu\left(\frac{\partial v_x}{\partial z} + \frac{\partial v_z}{\partial x}\right)\right] + \rho g_x \tag{8}$$

Equation 8 makes use of the continuity equation; analogous equations apply to y,z directions. In Equation 8, μ (dynamic viscosity; see Chapter 1, Section II) is the physical parameter interpreting the proportionality between the stress field and the velocity gradient field. A particular form of general momentum equations (Equation 8 and similar in the other directions), applicable to newtonian and incompressible (ρ = constant) fluids in isothermal processes (μ = constant), and which makes use of the continuity equation, is the set of Navier-Stokes equations:

$$\rho \frac{Dv_x}{Dt} = -\frac{\partial p}{\partial x} + \mu \nabla^2 v_x + \rho g_x \tag{9a}$$

$$\rho \frac{Dv_y}{Dt} = -\frac{\partial p}{\partial y} + \mu \nabla^2 v_y + \rho g_y \tag{9b}$$

$$\rho \frac{Dv_z}{Dt} = -\frac{\partial p}{\partial z} + \mu \nabla^2 v_z + \rho g_z \tag{9c}$$

If a newtonian, incompressible fluid also has a negligible dynamic viscosity (ideal fluid), momentum equations assume the following form in vectorial notation:

$$\rho \frac{D\vec{v}}{Dt} = -\text{grad } p + \rho \vec{g} \tag{10}$$

Equation 10 is known as the Euler equation.

C. Energy Equation

The energy conservation equation is obtained as a direct application of the first thermodynamic law, with the hypothesis that the control volume in the fluid system interacts energetically with the external environment only through exchanges of heat and work.

When analyzing the energy interactions, it is still necessary to separately analyze the various contributions, many of which are dependent on the velocity field, requiring its definition. In the evaluation of heat transfer, the radiative contributions have been disregarded in this work, due to their limited importance in "low" temperature situations, commonly encountered in applications of heat transfer with flowing fluids. Analogously, chemical, electrical, and nuclear interactions have been disregarded. Hypothesizing about the importance of such energies, it would be easy to include their contribution in the general form of energy conservation equation.

Energy balance applied to the control volume in a fluid system may be represented as (Figure 3): (energy variation in the control volume [c.v]) = + (energy entering the c.v.) − (energy leaving the c.v.) + (heat exchanged [+ = entering] with environment through conduction [with surrounding fluids or solids]) − (net work carried out by the fluid belonging to the c.v.). The convective contribution in the energy balance must include both the internal and the kinetic energy; the specific energy will be considered, therefore, as:

$$e = u + v^2/2 \tag{11}$$

The conductive contribution utilizes the well-known Fourier equation.

The term net work, carried out on the environment, includes the contribution of volume forces (i.e., gravity) and a contribution of surface forces.

In its general form, the energy conservation equation in cartesian coordinates may be represented as:

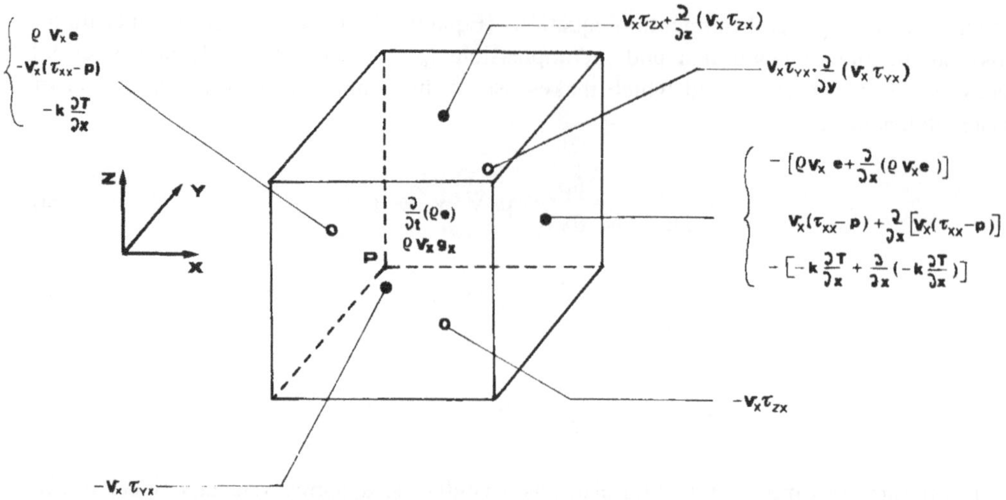

FIGURE 3. Reference control volume. Energy equation with a flow in x-axis direction.

$$\frac{\partial}{\partial t}(\rho e) = -\left[\frac{\partial}{\partial x}(\rho\, v_x\, e) + \frac{\partial}{\partial y}(\rho\, v_y\, e) + \frac{\partial}{\partial z}(\rho\, v_z\, e)\right] +$$

$$+ \left[\frac{\partial}{\partial x}\left(k\frac{\partial T}{\partial x}\right) + \frac{\partial}{\partial y}\left(k\frac{\partial T}{\partial y}\right) + \frac{\partial}{\partial z}\left(k\frac{\partial T}{\partial z}\right)\right] - \left[\frac{\partial}{\partial x}(pv_x) + \frac{\partial}{\partial y}(pv_y) +\right.$$

$$\left. + \frac{\partial}{\partial z}(pv_z)\right] + \left[\frac{\partial}{\partial x}(v_x\, \tau_{xx} + v_y\, \tau_{xy} + v_z\, \tau_{xz}) +\right.$$

$$+ \frac{\partial}{\partial y}(v_x\, \tau_{yx} + v_y\, \tau_{yy} + v_z\, \tau_{yz}) +$$

$$\left. + \frac{\partial}{\partial z}(v_x\, \tau_{zx} + v_y\, \tau_{zy} + v_z\, \tau_{zz})\right] + \rho(g_x\, v_x + g_y\, v_y + g_z\, v_z) \qquad (12)$$

or, using the substantial derivative notation, the continuity equation, the divergence and Laplace operators, as (see Appendix):

$$\rho\frac{De}{Dt} = \nabla(k\nabla T) + \rho\, \Sigma_i\, v_{x_i}\, g_{x_i} - \nabla(p\, \vec{v}) +$$

$$+ \Sigma_i\left[\frac{\partial}{\partial x_i}\Sigma_j\,(v_{x_j}\, \tau_{x_i x_j})\right] \qquad (13)$$

where: $i = 1, 2, 3$; $j = 1, 2, 3$; $x_1 \equiv x$; $x_2 \equiv y$; $x_3 \equiv z$. With the introduction of enthalpy (specific) $i = u + p/\rho$ and of total enthalpy (specific) $i' = i + v^2/2$, the energy equation (making use of the continuity equation) may be written as:

$$\rho\frac{Di'}{Dt} - \frac{\partial p}{\partial t} = \nabla(k\nabla T) + \rho\, \Sigma_i\, v_{x_i}\, g_{x_i} + \Sigma_i\left[\frac{\partial}{\partial x_i}\Sigma_j\,(v_{x_j}\, \tau_{x_i x_j})\right] \qquad (14)$$

In common applications, the energy conservation equation is used in forms which are obtained from Equation 14. Particular importance is assumed by the following forms, known respectively as the mechanical-energy (Equation 15) and the thermal-energy (Equation 16) conservation equations:

$$\rho \frac{D\left(\frac{v^2}{2}\right)}{Dt} + \Sigma_i \left(v_{x_i} \frac{\partial p}{\partial x_i}\right) = \Sigma_i v_{x_i} \left(\Sigma_j \frac{\partial \tau_{x_j x_i}}{\partial x_j}\right) + \Sigma_i \rho v_{x_i} \cdot g_{x_i} \tag{15}$$

$$\rho \frac{Du}{Dt} + p \nabla \vec{v} = \nabla(k \nabla T) + \Sigma_i \left[\Sigma_j \tau_{x_i x_j} \frac{\partial v_{x_j}}{\partial x_i}\right] \tag{16}$$

The viscous dissipation term $\Sigma_i (\Sigma_j \tau_{x_i x_j} \partial V_{x_j}/\partial x_i)$ is often mentioned as a dissipation function, ϕ. In the particular case of a newtonian fluid, using the constitutive relationships given by the Stokes law:

$$\phi = \mu\left\{2\left[\left(\frac{\partial v_x}{\partial x}\right)^2 + \left(\frac{\partial v_y}{\partial y}\right)^2 + \left(\frac{\partial v_z}{\partial z}\right)^2\right] + \left(\frac{\partial v_y}{\partial x} + \frac{\partial v_x}{\partial y}\right)^2 + \left(\frac{\partial v_z}{\partial y} + \frac{\partial v_y}{\partial z}\right)^2 + \right.$$
$$\left. + \left(\frac{\partial v_x}{\partial z} + \frac{\partial v_z}{\partial x}\right)^2 - \frac{2}{3}\left(\frac{\partial v_x}{\partial x} + \frac{\partial v_y}{\partial y} + \frac{\partial v_z}{\partial z}\right)^2\right\} \tag{17}$$

A further form of the energy equation, of great interest in applications, is obtained, introducing in Equation 16 the dependence on temperature:

$$\rho c_v \frac{DT}{Dt} = -T\left(\frac{\partial p}{\partial T}\right)_v \nabla \vec{v} + \nabla(k \nabla T) + \phi \tag{18}$$

Equation 18 is often specialized to particular applications and shows how fluid temperature varies as an effect of thermal conduction, compression, and viscous dissipation:

- For an ideal gas with a constant thermal conductivity and a negligible viscous dissipation factor, ϕ, being $(\partial p/\partial T_V) = p/T$, it is

$$\rho c_v \frac{DT}{Dt} = k \nabla^2 T - p \nabla \vec{v} \tag{19}$$

- For a fluid at constant pressure and thermal conductivity, neglecting the viscous dissipation, being

$$du = -p\, d\, v + c_p\, d\, T \tag{20}$$

and being (from Equation 20 and the continuity equation):

$$\rho \frac{Du}{Dt} = -p \rho \frac{Dv}{Dt} + \rho c_p \frac{DT}{Dt} = -p \nabla \vec{v} + \rho c_p \frac{DT}{Dt} \tag{21}$$

it is

$$\rho c_p \frac{DT}{Dt} = k \nabla^2 T \tag{22}$$

where v is the specific volume.

- For a liquid (ρ independent of T), assuming constant thermal conductivity and neglecting the viscous dissipation, it is

$$\rho\, c_p \frac{DT}{Dt} = k\, \nabla^2 T \tag{23}$$

- For a solid (here reported for completeness), being ρ = const, v = 0 velocity; $\phi = 0$, the Fourier equation is obtained (in the case of a constant thermal conductivity):

$$\rho\, c_p \frac{\partial T}{\partial t} = k\, \nabla^2 T \tag{24}$$

III. TURBULENT FLOWS

The analysis of turbulence phenomena began in the late 19th century with the studies and classical treatments of Boussinesque and Reynolds; recently, this analysis received a very important push from the modern aircraft and space industries.[2-4]

The preeminent characteristic of turbulent flows is the continuous tendency to mixing of the fluid particles, which leads (even in stationary conditions, under a macroscopic point of view) to fluid particles which, in time, change the direction and/or the modulus of their velocity chaotically and incoherently with respect to the macroscopic flow direction. The nature and mode of the mixing have been interpreted considering the formation of a cascade of vortexes (Von Kármán vortexes), with dimensions decreasing and the final formation of very small vortexes, where the dissipation of mechanical energy into heat occurs.

A representation of turbulent flow characteristics, using the same parameters as in the analysis of laminar flows (local velocities, local pressure, etc.), involves great difficulty owing to the presence of the chaotic and very rapid variations of such parameters, whatever dimensional representation scale has been chosen. Furthermore, the punctual representation of the fluid volume would lead, due to the impossibility of obtaining analytical solutions, to a discretization in a great number of meshes, in order to follow the continuous local fluctuations by means of discretized variations. Such representation, quite costly from a computational point of view, could not indicate the overall relevant macroscopic aspects of the flow.

In order to achieve an effective representation of a turbulent flow, the first step is the definition of the most suitable representation scale (in time and in space) and, therefore, that of the most significant parameters themselves (local and instantaneous values of velocities and pressure are not suitable). In this regard, it has been noted that in the applications of usual interest, within the turbulent flow, very fast local fluctuations coexist with variations slower in time. Some of the most employed turbulence representation techniques (for a flow internal or external to a solid) exploit this behavior and define, as representative variables of the fluid status, the average values and the fluctuant components of the describing parameters already used for the laminar flow (pressure, temperature, and velocities). The use of the average values of the physical parameters allows a first comparison between an "overall" behavior of the turbulent flow (defined by the average values of the representative parameters) and the behavior of the laminar flow (using the corresponding parameters). From such comparisons, it may be pointed out that with equal values of the average parameters (in laminar flows the average values are coincident with the local values), greater pressure drops occur in turbulent flows due to the vorticous situation. This may be interpreted as if in turbulence conditions the fluid were more viscous; the entity of this viscosity increment is a function of the pressure and of the specific flow situation. Analogously, it may be noted that with the same thermal gradient imposed as a boundary condition to the fluid, the heat flux transmitted in a turbulent flow is greater than the corresponding flux in a laminar flow. This may be interpreted as though the fluid had a greater thermal conductivity; this increment is also a function of the turbulence level.

The possibility of such analogies with the laminar flow conditions represents a consistent advantage, because it allows the reduction of the problem of representing the chaotic flow in turbulent conditions to the definition of simple parameters as the "turbulent viscosity" and the "turbulent conductivity". Clearly, they do not have the same meaning as the corresponding ones in laminar flow; in such a case viscosity and thermal conductivity are characteristic parameters of the molecular dimensional scale and have a well-defined physical meaning (they are physical properties; see Chapter 1). In the case of the turbulent viscosity and thermal conductivity, the reference scale is relative to greater dimensions (turbulence vortexes) and the definition only has a practical meaning (their values depend on the local turbulent level which, *ceteris paribus,* also depends on the specific geometric configuration).

On the basis of the mentioned approach, which makes use of the average and fluctuant values of the main fluidodynamic parameters (velocities, pressure, temperature, etc.), if the variations of density are small, the conservation equations with a turbulent flow may be written as (overlined magnitudes represent the average components, while the variables with apices are the fluctuant components; see Equation 64):

- Continuity

$$\frac{\partial \rho}{\partial t} + \frac{\partial}{\partial x}(\rho\,\overline{v}_x) + \frac{\partial}{\partial y}(\rho\,\overline{v}_y) + \frac{\partial}{\partial z}(\rho\,\overline{v}_z) = 0 \tag{25}$$

- Momentum

$$\rho\,\frac{D\overline{v}_x}{Dt} = -\frac{\partial \overline{p}}{\partial x} + \mu\,\nabla^2\,\overline{v}_x - \Sigma_i\,\frac{\partial}{\partial x_i}(\rho\,\overline{v'_x\,v'_{x_i}}) + \rho\,g_x \tag{26}$$

for a constant-property fluid, in direction x (analogous equations are valid in y and z directions); $T^t_{xx_i} = -\rho\,\overline{v'_x\,v'_{x_i}}$ is a component of the Reynolds stresses tensor (see Section III.A):

- Energy

$$\rho\,c_p\,\frac{D\overline{T}}{Dt} = \nabla(k\nabla\overline{T}) - \Sigma_i\,\frac{\partial}{\partial x_i}(\rho\,c_p\,\overline{v'_{x_i}\,T'}) \tag{27}$$

for a low-velocity flow and negligible volume and viscous forces; $q''^t_{xi} = \rho\,c_p\,\overline{v'_{x_i}\,T'}$ is the turbulent energy flux in direction x_i.

The definition of a greater number of unknown parameters (both the average and the fluctuant components) does not allow the solution of problems using only the indicated equations. There is the necessity of further relationships, containing some "physical" information on the fluid system (characterizing the dynamic structure of vortexes and mechanical energy dissipation modes in them), in order to equate the total number of equations (or relationships) and that of the unknown parameters. In the definition of such further relationships, turbulence models do consist.

Before analyzing the turbulence models, it must be emphasized that the Navier-Stokes equations (momentum equations such as Equation 9) should also describe turbulent flows, but they are advantageously used only in applications with laminar flows. Their solution, in turbulent flow, would result in the actual path followed by the fluid particles as an effect of the irregular random and global motions. Even utilizing the present powerful computers, a solution is not practically possible. Therefore, it is necessary to obtain information on time-averaged quantities, for example, following the Reynolds' procedure, which leads to Equations 25, 26, and 27, and using the conventionally defined fluctuant parameters, whose representation and proposed dependence on the averaged values is arbitrary.

A. Turbulence Models

For the solution of the conservation equations, some further relationships are necessary, in particular for the evaluation of the fluctuant-component terms of Equation 26 (the "stresses" due to the fluctuant components of the velocities) and of Equation 27.

1. Momentum

The fluctuant component terms of Equation 26 correspond to further stresses inside the fluid, due to the turbulence (Reynolds stresses). Using a matricial representation (matrix of Stokes stresses), the Reynolds stresses tensor is defined as:

$$\begin{pmatrix} \tau^t_{xx} & \tau^t_{xy} & \tau^t_{xz} \\ \tau^t_{xy} & \tau^t_{yy} & \tau^t_{yz} \\ \tau^t_{xz} & \tau^t_{yz} & \tau^t_{zz} \end{pmatrix} = -\begin{pmatrix} \rho\overline{v_x'^2} & \rho\overline{v_x' v_y'} & \rho\overline{v_x' v_z'} \\ \rho\overline{v_x' v_y'} & \rho\overline{v_y'^2} & \rho\overline{v_y' v_z'} \\ \rho\overline{v_x' v_z'} & \rho\overline{v_y' v_z'} & \rho\overline{v_z'^2} \end{pmatrix} \tag{28}$$

For the evaluation of the Reynolds stresses, several models or approaches have been proposed (i.e., k-model, k-ϵ model), whose treatment is generally rather complex. For a complete analysis, see Reference 4.

The different techniques proposed differ from one another in the complexity and approximation of the treatment itself (the proposed relationships, relating the fluctuant components to the average are empirical in nature) and in the specific applications with which their use is compatible.

One of the most significant approaches is the "turbulent viscosity model", initially proposed by Boussinesque, based on the assumption of an analogy between the momentum transport mechanism in laminar and in turbulent flows. In analogy with Newton's law, $\tau_{xy} = \mu \partial v_x/\partial y$ (valid for a two-dimensional laminar flow with the main direction parallel to axis x), may be written:

$$\tau^t_{xy} = \mu_t\, \partial\overline{v}_x/\partial y \tag{29}$$

The turbulent viscosity μ_t may be introduced by means of the relationship (from Equations 28 and 29:

$$-\overline{v_{x_i}' v_{x_j}'} = \frac{\mu_t}{\rho}\,(\partial\overline{v}_{x_i}/\partial x_j + \partial\overline{v}_{x_j}/\partial x_i) \tag{30}$$

The definition of the parameter μ_t requires the evaluation of the link with flow characteristics; this link is not exactly definable and many correlations have been proposed, differing both in their formulation and in the applicability field.

The simpliest model for the definition of the turbulent viscosity was introduced by Prandtl, who proposed putting the fluctuant component of the viscosity in relation to the average velocity gradient multiplied by a distance, l_m, called "mixing length", which was assumed to be independent of the average value of the velocity and empirically evaluated. It is (in a two-dimensional flow):

$$v'_{x_i} = l_m(\partial \bar{v}_{x_i}/\partial x_j + \partial \bar{v}_{x_j}/\partial x_i) \quad (31)$$

It may be supposed that v'_{x_j} is of the same order of magnitude as v'_{x_i} (circular trajectories of vortexes); therefore, as a consequence of Equation 30, the relationship existing between μ_t and l_m may be written:

$$\mu_t/\rho = l_m^2|\partial \bar{v}_{x_i}/\partial x_j + \partial \bar{v}_{x_j}/\partial x_i| \quad (32)$$

For the generic Reynolds stress $\tau^t_{x_ix_j}$, it may be written (see Equations 28 and 31):

$$\tau^t_{x_ix_j} = \rho\, l_m^2 \left|\frac{\partial \bar{v}_{x_i}}{\partial x_j} + \frac{\partial \bar{v}_{x_j}}{\partial x_i}\right| \cdot \left(\frac{\partial \bar{v}_{x_i}}{\partial x_j} + \frac{\partial \bar{v}_{x_j}}{\partial x_i}\right) \quad (33)$$

where the sign of τ^t_{xixj} is consequent to that of the velocity gradient.

The mixing length evaluation allows the closure of the momentum equations. The mixing length is a function of the position in the flow. As an example, in a two-dimensional flow inside a channel, Prandtl suggests $l_m = C \cdot y$, where y is the distance from the wall (experimentally, it has been evaluated $C \simeq 0.4$).

The Prandtl theory is applied with success in the study of turbulent flows in proximity to solid walls (channels, slabs) and in problems of "free turbulence" (jets, atmospheric turbulence). It is not suitable in proximity to points where the velocity gradient becomes zero, also leading (according to Equation 33), to a small value of stress $\tau_{x_ix_j}$; this condition has experimentally proven to be inexact.

The Prandtl method has been subjected to improvements, in order to increase its range of applicability. On the basis of dimensional considerations, Von Kármán proposed the following relationship for the mixing length:

$$l_m = \chi|(\partial \bar{v}_x/\partial y)/(\partial^2 \bar{v}_x/\partial y^2)| \quad (34)$$

from which the expression of the Reynolds stress in two-dimensional problems may be obtained as:

$$\tau^t_{xy} = \rho\, \chi^2 \left|\frac{\left(\frac{\partial \bar{v}_x}{\partial y}\right)^3}{\left(\frac{\partial^2 \bar{v}_x}{\partial y^2}\right)^2}\right| \cdot \left(\frac{\partial \bar{v}_x}{\partial_y}\right) \quad (35)$$

where $\chi = 0.36$ to 0.4, from velocity profiles in tubes.

More complex models have been proposed to represent the dependence of the turbulent viscosity from the flow conditions; a detailed treatment of such models is seen in Reference 4.

2. *Energy*

Following the definition of the fluctuant component of the fluid temperature, an analogy may be assumed between the turbulent heat transport and the laminar heat conduction transport introducing the "turbulent conductivity". Using the Fourier law, the turbulent conductivity may be introduced, such that:

$$\rho\, c_p\, \overline{v'_{x_i} T'} = k_t\, \partial \bar{T}/\partial x_i \quad (36)$$

Consequently, a turbulent diffusivity, α_t, may be defined as:

$$\alpha_t = k_t/(\rho\, c_p) = \overline{v'_{x_i}\, T'}/(\partial \overline{T}/\partial x_i) \qquad (37)$$

It is important to emphasize that both μ_t and k_t are not thermodynamic properties of the fluid, but are "apparent" parameters, depending on the turbulence level and varying with it. A turbulent Prandtl Number may be defined as:

$$Pr_t = \mu_t/(\rho\, \alpha_t) \qquad (38)$$

Even if this quantity is variable in the function of the flow conditions, it may be often considered as a constant. The Reynolds' analogy refers to that situation, considering a similar mechanism of turbulent exchange for the transfer of momentum and heat.

As correlations proposed to relate μ_t to the flow characteristics are empirical and somehow arbitrary, analogous considerations apply to parameter k_t (or α_t). One of the most common models assumes $Pr_t = 1$ and defines $\alpha_t = \mu_t/\rho$, allowing — through the correlation expressing μ_t — the solution of Equation 27.

IV. TWO-PHASE FLOWS

It has been noted that the solution of conservation equations for a single-phase turbulent flow requires some approximations and the definition of empirical relationships (closure models). In the situation of two-phase flows, the contemporary presence of two phases interacting with each other in energy and momentum exchanges, determines a considerable complication of the analytical problem of solving conservation equations. This is a fact owing to the need of describing not one, but two turbulent flows, and owing to the continuous variations of the most representative variables.

Historically, the first approaches have considered the two-phase flow as a single-phase flow of adequate characteristics (homogeneous models), but the results are acceptable only in situations of thermodynamic equilibrium, dispersion of one phase in the other, and absence of thermodynamic transformations in space or in time domain.

Subsequently and alternatively, more complicated models have been proposed, considering a separation of the two phases. This approach requires a separate analysis of each phase, the application of conservation equations to each of them, and the definition of mass, momentum, and energy transport between the two phases. The very difficult problem of correctly evaluating the phases' interactions is complicated by the mathematical difficulty of solving rather complex differential equation systems. The present tendency is, as a consequence, that of starting from instantaneous and punctual equations for each phase, subjecting these equations to averaging procedures in time and/or space domain, applying the simplifications finalized at pointing out and using only the most relevant and representative parameters, and researching the most suitable "closure" relationships in function of the specific problem.

It is clear that in the averaging stage some information is lost, especially regarding the parameter gradients at the interface or close to the channel wall. In the choice of the averaging techniques that must filter the fluctuations of parameters, they must prove not to affect the space and time oscillations of the "average" values.[5-8] In the following paragraphs, the symbol "< >" is used to indicate a spatial averaging, and the symbol "—" to indicate a time averaging.

A. Instantaneous and Punctual Form

Conservation equations may be obtained in their most general form (instantaneous, punc-

Table 1
MEANING OF SYMBOLS

Equation	ψ_k	$\vec{J}_k$	ϕ_k
Mass	1	0	0
Momentum	$\vec{v}_k$	$-\vec{\tau}^{\bullet}_k = +\vec{p}_k - \vec{\tau}_k$	$\vec{F}$
Energy	$\frac{1}{2} v_k^2 + u_k$	$\vec{q}''_k - \vec{v}_k \vec{\tau}^{\bullet}_k$	$\vec{F} \cdot \vec{v}_k + \frac{\dot{Q}_k}{\rho_k}$

tual form) starting from mass, momentum, and energy balances for each phase, with reference to a control volume, then analytically modifying the integral balances (i.e., applying the Gauss theorem, etc.), in order to obtain the local and instantaneous differential equations for each phase.

In this paragraph only the ''final'' form of the equations is reported[5]:

- Continuity

$$\partial \rho_k/\partial t + \nabla(\rho_k \vec{v}_k) = 0 \tag{39}$$

- Momentum

$$\frac{\partial(\rho_k \vec{v}_k)}{\partial t} + \nabla(\rho_k \cdot \vec{v}_k \cdot \vec{v}_k) - \rho_k \vec{F} - \nabla \vec{\tau}^{\bullet} = 0 \tag{40}$$

- Energy

$$\frac{\partial}{\partial t}\left[\rho_k\left(u_k + \frac{1}{2} v_k^2\right)\right] + \nabla\left[\rho_k \left(\frac{1}{2} v_k^2 + u_k\right) v_k\right] - \dot{Q}_k - \rho_k \vec{F} \cdot \vec{v}_k$$
$$- \nabla\left(\vec{\tau}^{*}_k \cdot \vec{v}_k\right) + \nabla q''_k = 0 \tag{41}$$

where $\vec{F}$ is the external force per unit mass [N/kg], $\vec{\tau}^{\bullet}$ is the stress tensor ($-\vec{p} + \dot{\vec{\tau}}$; [N/m²]), k indicates each phase (k = l; v), $\vec{q}''_k$ is the heat flux vector [J/(m²s)], and $\dot{Q}_k$ is the amount of heat produced in the phase k in the unit of volume and of time [J/(m³s)].

The general form of instantaneous and punctual conservation equation is

$$\partial(\rho_k \psi_k)/\partial t + \nabla(\rho_k \psi_k \vec{v}_k) + \nabla \vec{J}_k - \rho_k \phi_k = 0 \tag{42}$$

where ψ_k, $\vec{J}_k$, and ϕ_k are symbols whose meaning is contained within Table 1.

Particular care must be addressed to boundary conditions.[5] Interface congruence conservation equations are of the kind:

$$\sum_{k=l,v} (G_k \psi_k + \vec{n}_k \cdot \vec{J}_k) = 0 \tag{43}$$

where

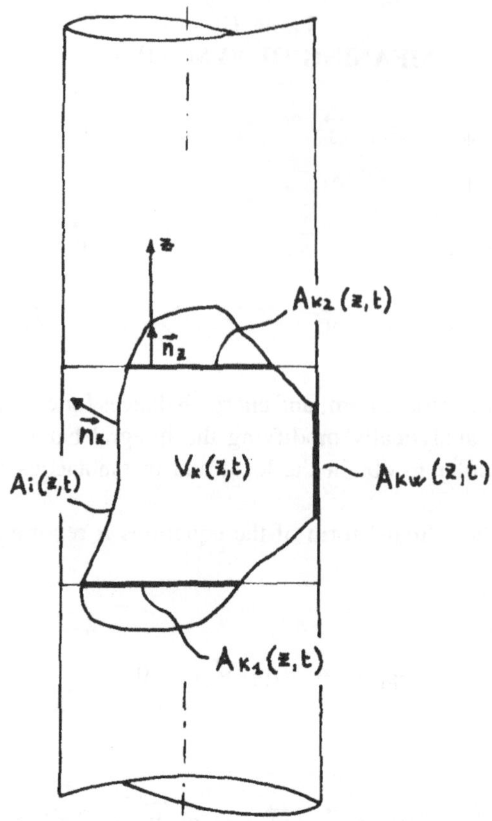

FIGURE 4. Control volume with a two-phase flow.

$$G_k = \rho_k(\vec{v}_k - \vec{v}_i) \cdot \vec{n}_k \tag{44}$$

is the mass flux of the phase k through the interface i (whose velocity is $\vec{v}_i$).

B. Instantaneous Equations Averaged in Space

For simplicity, a one-dimensional geometry is considered. In particular, a fixed channel is considered, with axis z (unit vector $\vec{n}_z$), in which a volume is cut by two cross-sectional planes located at a distance z (Figure 4). V_k (z,t) is the volume limited by A_{K1} and A_{K2}; A_i being the portions of the interface and A_{kw} being that of the wall enclosed between the two cross-sectional planes. The unit vector normal to the interface and directed away from phase k is denoted by $\vec{n}_k$.

According to Equation 42, ψ_k is the parameter to be conserved, $\vec{J}_k$ is the flux, and ϕ_k the source term of ψ_k.

With the following definitions:

$$\langle f_k \rangle = (1/V_k) \int_{V_k} f_k(z, t)\, d\, V$$

(volumetric average value of the generic f_k function)*

* At the surface or at the wall: $\langle f_k \rangle_{i,w} = \dfrac{1}{V_k} \displaystyle\int_{A_{i,w}} f_k(z, t)\, d\, A.$

$$\alpha_k = V_k/V$$

the conservation equations, averaged in space (volume), may be written assuming surfaces A_{k1} and A_{k2} as fixed:[5,6]

- Mass

$$\frac{\partial}{\partial t}(\alpha_k \langle \rho_k \rangle) + \frac{\partial}{\partial z}(\alpha_k \langle \rho_k v_{z,k} \rangle) = -\langle G_k \rangle_i \tag{45}$$

with $v_{z,k}$ the component in the direction z, of the velocity vector $\vec{v}_k$.

In general, the volume-average value of the interfacial mass transfer $\langle G_k \rangle_i$ is not known *a priori* and must be evaluated with adequate correlations.

- Momentum

$$\begin{aligned}
&\frac{\partial}{\partial t}(\alpha_k \langle \rho_k v_{z,k} \rangle) + \frac{\partial}{\partial z}(\alpha_k \langle \rho_k v_{z,k}^2 \rangle) + \alpha_k \frac{\partial \langle p_k \rangle}{\partial z} + \\
&- \frac{\partial}{\partial z}(\alpha_k \langle \tau_{zz,k} \rangle) - \Delta p_{k,i} \frac{\partial \alpha_k}{\partial z} = \alpha_k \langle \rho_k F_{z,k} \rangle_i + \\
&- \langle G_k v_{z,k} \rangle_i - \langle \vec{n}_k \cdot \vec{n}_z \Delta p'_{k,i} \rangle_i + \langle \vec{n}_k \cdot \vec{\tau}_z \rangle_i + \langle \vec{n}_{k,w} \cdot \vec{\tau}_z \rangle_w
\end{aligned} \tag{46}$$

where $\Delta p_{k,i} = \langle p_{ki} \rangle - \langle p_k \rangle$ (difference between the bulk and the interfacial pressure) and

$$\Delta p'_{k,i} = p_k - \langle p_{k,i} \rangle,$$

$$p_k = \langle p_k \rangle + \Delta p_{k,i} + \Delta p'_{k,i}$$

In Equation 46, the term containing $\Delta p'_{k,i}$ on the right-hand side, is the term that leads to the "apparent mass" effect for inviscid flows. It may be calculated if the shape of the interface is known *a priori*. Otherwise (as generally happens), semi-empirical correlations are necessary[6]:

- Energy

$$\begin{aligned}
&\frac{\partial}{\partial t}\left[\alpha_k \langle \rho_k \left(i_k + \frac{v_k^2}{2}\right) \rangle\right] + \frac{\partial}{\partial z}\left[\alpha_k \langle \rho_k v_k \left(i_k + \frac{v_k^2}{2}\right) \rangle\right] + \\
&- \alpha_k \frac{\partial \langle p_k \rangle}{\partial t} + \Delta p_{k,i} \frac{\partial \alpha_k}{\partial t} + \frac{\partial}{\partial z} \alpha_k \langle q_{z,k} \rangle - \frac{\partial}{\partial_z} [\alpha_k \langle \vec{n}_z \cdot \\
&\cdot (\vec{\tau}_k \cdot \vec{v}_k) \rangle] = -\langle \left[G_k \left(i_k + \frac{v_k^2}{2}\right) + \vec{n}_k \cdot \vec{v}_i \Delta p'_{k,i} + \vec{n}_k \cdot \vec{q}''_k + \\
&- \vec{n}_k \cdot (\vec{v}_k \cdot \vec{\tau}_k) \right] \rangle_i - \langle (\vec{n}_k \cdot \vec{q}''_k) \rangle_w + \alpha_k \langle (\rho_k \vec{v}_k \cdot \vec{F}_k + Q_k) \rangle
\end{aligned} \tag{47}$$

Many investigators assume that $p = p_l = p_v = p_{l,i} = p_{v,i}$. The equations presented are so simplified because terms including $\Delta p_{k,i}$ and $\Delta p'_{k,i}$ disappear. Futhermore, p_k is replaced by

p. In addition to flowing phases conservation equations, the interface congruence conservation equations may also be averaged in space:

- Mass

$$\langle G_l \rangle_i = -\langle G_v \rangle_i \tag{48}$$

- Momentum

$$[\langle p_{l,i} \rangle - \langle p_{v,i} \rangle] \frac{\partial \alpha_l}{\partial z} = \langle G_l(v_{z,l} - v_{z,v}) + (\Delta p'_{l,i} - \Delta p'_{v,i}) + $$

$$- \vec{n}_l(\vec{\tau}_{z,l} - \vec{\tau}_{z,v}) \rangle_i \tag{49}$$

- Energy

$$[\langle p_{l,i} \rangle - \langle p_{v,i} \rangle] \frac{\partial \alpha_l}{\partial t} = \langle \vec{n}_l \cdot [(\vec{v}_l \cdot \vec{\tau}_l) - (\vec{v}_v \cdot \vec{\tau}_v)] +$$

$$- \vec{n}_l \cdot (\vec{q}''_l - \vec{q}''_v) - \vec{n}_l \cdot \vec{v}_l(\Delta p'_{l,i} - \Delta p'_{v,i}) - G_l\left[\left(i_l + \frac{v_l^2}{2}\right) + \right.$$

$$\left. - \left(i_v + \frac{v_v^2}{2}\right)\right] \rangle_i \tag{50}$$

In general, the terms $\Delta p'_{k,i}$ could be very small with respect to $\vec{q}''_k$ and i_k.

In the above relationships, if the flow is characterized by a negligible mass transfer through the interface and neglecting the surface tension, the difference $< p_{l,i} > - < p_{v,i} >$ is usually assumed to vanish. In the opposite situation, with large gradients of α, the term containing $\Delta p'_{k,i}$ becomes small (fast condensation or vaporization).

C. Punctual Equations Averaged in Time

In order to describe a flow in transient conditions with the use of a discretization, a time averaging may be performed for the punctual and instantaneous equations (see Section IV.A).

The time interval $[t - t^*/2; t + t^*/2]$ must be chosen sufficiently large with respect to the turbulent fluctuations, but sufficiently small with respect to the macroscopic fluctuations which characterize the transient phenomenon.

Phase k passes through a specific point intermittently; a function f_k (each subinterval is continuous in which the phase k passes through the point) may be associated to the phase k (Figure 5).

If $[t_k^*]$ is the subset of residence time intervals of phase k belonging to the interval $[t^*]$ and t_k^* is the cumulative residence time of phase k in the point, within the interval $[t^*]$, a fraction of residence time r_k may be defined as:

$$r_k = t_k^*/t^* \tag{51}$$

A density-phase function $X_k(\vec{x},t)$ may also be introduced, defined as:

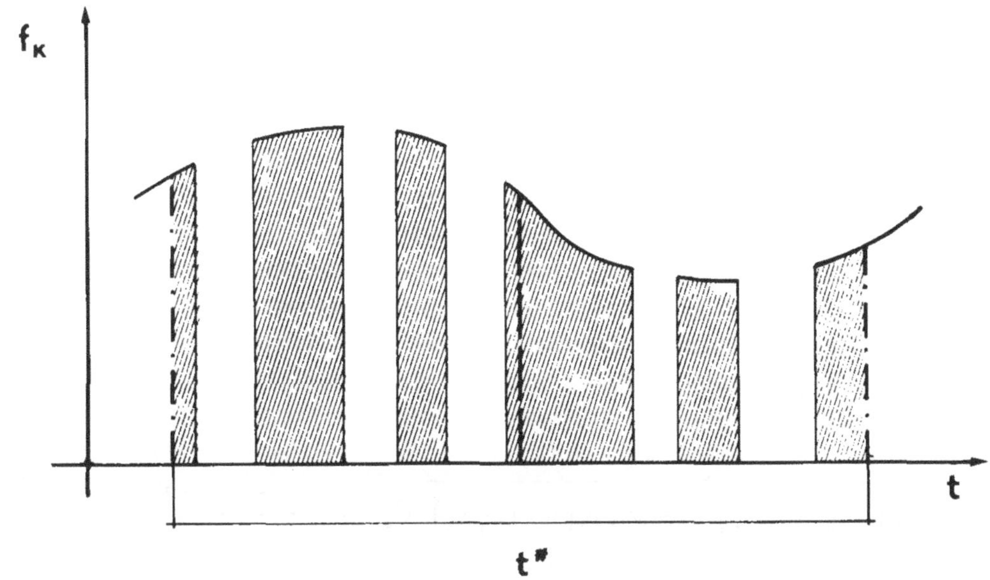

FIGURE 5. Succession of phases through a point.

$$X_k(\vec{x}, t) = \begin{cases} 1 & \text{if the point corresponding to vector } \vec{x} \text{ belongs to phase k} \\ 0 & \text{if the point corresponding to vector } \vec{x} \text{ does not belong to phase k} \end{cases}$$

The fraction of residence time function r_k is then:

$$r_k(\vec{x}, t) = \frac{t_k^*}{t^*} = \frac{1}{t^*} \int_{[t^*]} X_k(\vec{x}, t)\, d\, t = \overline{X}_k(\vec{x}, t) \tag{52}$$

The time-averaged value of function f_k in the interval $[t_k^*]$ is

$$\overline{f_k}^X = \frac{1}{t_k^*} \int_{[t_k^*]} f_k\, d\, t = \frac{\frac{1}{t^*} \int_{[t^*]} X_k\, f_k\, d\, t}{\frac{1}{t^*} \int_{[t^*]} X_k\, d\, t} = \frac{\overline{X_k\, f_k}}{\overline{X_k}} \tag{53}$$

The time-averaged value of f_k is equal to the average value of f_k weighted on X_k, in the interval $[t^*]$. On the basis of such definitions, through analytical transformations, the conservation equations in punctual form, averaged in time, are[5]

- Mass

$$\frac{\partial}{\partial t}\, \overline{r_k\, \rho_k}^X + \nabla\, \overline{r_k\, \rho_k\, \vec{v}_k}^X = -\Sigma_j\, l_j^{-1} (G_k)_j \tag{54}$$

where $l_j = t^* |\vec{v}_i \cdot \vec{n}_k|$ is called "specific length" and j is relative to the j th interface passing through x in the interval [t*]:

- Momentum

$$\frac{\partial}{\partial t} \overline{r_k \cdot \rho_k \cdot \vec{v}_k}^X + \nabla\, \overline{r_k \cdot \rho_k \cdot \vec{v}_k \cdot \vec{v}_k}^X - \nabla\, \overline{r_k \vec{\tau}_k^*}^X - \overline{r_k \rho_k \vec{F}_k}^X =$$

$$= -\Sigma_j \frac{(G_k \vec{v}_k - \vec{\tau}_k^* \vec{n}_k)_j}{l_j} \tag{55}$$

- Energy

$$\frac{\partial}{\partial t} \overline{r_k \rho_k \left(u_k + \frac{1}{2} v_k^2\right)}^X + \nabla \overline{r_k \rho_k \left(u_k + \frac{1}{2} v_k^2\right) \vec{v}_k}^X + \nabla \overline{r_k \vec{\tau}_k^* \cdot \vec{v}_k}^X +$$

$$+ \nabla \overline{r_k \vec{q}_k''}^X - \overline{r_k \rho_k \vec{F}_k \cdot \vec{v}_k}^X - \overline{r_k \dot{Q}_k}^X =$$

$$= -\Sigma_j \frac{\left[G_k\left(u_k + \frac{1}{2} v_k^2\right) - (\vec{\tau}_k^* \cdot \vec{v}_k) \cdot \vec{n}_k + \vec{q}_k'' \cdot \vec{n}_k\right]_j}{l_j} \tag{56}$$

D. Equations Averaged in Time and Space

If f_k is a function associated to the phase as a consequence of Equations 52 and 53, the following definition may be written:

$$\langle \overline{r_k f_k}^X \rangle = \langle \overline{X_k \cdot f_k} \rangle = \frac{1}{V} \int_V \left(\frac{1}{t^*} \int_{[t^*]} X_k f_k \, d\, t\right) d\, V \tag{57}$$

where the spatial averaging is considered in the control volume.

The fundamental relationship

$$\langle \overline{r_k f_k}^X \rangle = \overline{\alpha_k \langle f_k \rangle} \tag{58}$$

may be demonstrated[5] where:

$$\alpha_k = V_k/V$$

Equation 58 expresses the equivalence of the two possible procedures of compound averaging (in space and time), starting from punctual and time-averaged parameters (r_k and $\overline{f}_k^x$) or from instantaneous and volume-averaged parameters (α_k and $<f_k>$). The two procedures of averaging are alternately more or less convenient in the function of the actual flow regime (bubble, plug, slug, etc.; see Chapter 3).

A particular situation is obtained when $f_k = 1$, being $<r_k> = \overline{\alpha}_k$. If a generic conservation (balance) equation, punctual and averaged in time, is averaged in space, it is obtained as:

$$\frac{\partial}{\partial t}\overline{\langle r_k\, \rho_k\, \psi_k\rangle}^X + \frac{\partial}{\partial z}\overline{\langle r_k\, \rho_k\, \psi_k\, \vec{v}_k \cdot \vec{n}_z\rangle}^X + \frac{\partial}{\partial z}\overline{\langle r_k\, \vec{n}_z \cdot \vec{J}_k\rangle}^X +$$

$$-\overline{\langle r_k\, \rho_k\, \phi_k\rangle}^X = -\langle \Sigma_j \frac{(G_k\, \psi_k + \vec{J}_k \cdot \vec{n}_k)_j}{l_j}\rangle - \frac{1}{V}\int_{A_{k,w(z,t)}} \vec{n}_k \cdot \vec{J}_k\, d\, A \qquad (59)$$

In turn, if a generic conservation equation, instantaneous and averaged in space, is averaged in time, it is obtained as:

$$\frac{\partial}{\partial t}\alpha_k\,\langle\rho_k\,\psi_k\rangle + \frac{\partial}{\partial t}\overline{\alpha_k\,\langle\vec{n}_z(\rho_k\,\psi_k\,\vec{v}_k)\rangle} + \frac{\partial}{\partial z}\alpha_k\,\langle\vec{n}_k \cdot \vec{J}_k\rangle +$$

$$-\overline{\alpha_k\,\langle\rho_k\,\phi_k\rangle} = -\frac{1}{V}\int_{A_j(z,t)}(G_k\,\psi_k + \vec{n}_k\,\vec{J}_k)\,d\,A - \frac{1}{V}\int_{A_{k,w(z,t)}}\vec{n}_k \cdot \vec{J}_k\,d\,A \qquad (60)$$

It is possible to demonstrate that Equations 59 and 60 are identical. Through Table 1, the conservation equations for mass (continuity), momentum, and energy may be obtained, averaged in time and space.

E. Simplified Models

The use of conservation equations in their general forms (see Sections IV. A, B, C, and D) pertains only to very particular applications.

The use of simplified models for the analysis of two-phase mixtures is more usual; with simplified hypotheses on the interactions between the phases, they lead to generally very simple descriptive relationships. Such models have been developed with reference to quite particular geometries (i.e., tubular ducts) for which, in general, they furnish good results. A first class of models has been developed on the basis of particular hypothesized velocity profiles, function of the channel geometries. Such a class includes the one-dimensional homogeneous model, the Bankoff model,[9] the Wallis model,[10] and the Zuber-Findlay model.[11] In this chapter, only the homogeneous one-dimensional model and the Zuber-Findlay model are reported, the former being more suitable for quick and rough evaluations, while the latter is suitable to more accurate applications.

A second class of models (separated-flows models) separately describes the two phases and their interaction (in terms of mass, momentum, and energy transfer); such interactions are treated through simplifying hypotheses and on the basis of empirical correlations. These models provide more accurate results only if the empirical correlations describing the mechanical and thermal interactions between the phases are very accurate; they generally require the use of numerical computers.

1. Homogeneous One-Dimensional (One-Velocity) Model

A homogeneous dispersion of one phase into the other is assumed. The same velocity for the two phases may be hypothesized, together with the thermal equilibrium between them. With reference to a one-dimensional geometry, the conservation equations are written:

- Mass

$$\partial\rho_{mix}/\partial t + \partial(\rho_{mix}\, v_z)/\partial z = 0 \qquad (61)$$

- Momentum

$$\frac{\partial}{\partial t}(\rho_{mix} v_z) + \frac{\partial}{\partial z}(\rho_{mix} v_z^2) - \rho_{mix} F_z + \frac{\partial}{\partial z} p - \frac{\partial}{\partial z}\tau_z = 0 \tag{62}$$

where F_z is the z-component of the external force per unit mass (i.e., gravity force) and τ_z the shear stress in z direction.

- Energy

$$\frac{\partial}{\partial t}\left[\rho_{mix}\left(u + \frac{v_z^2}{2}\right)\right] + \frac{\partial}{\partial z}\left[\rho_{mix} v_z\left(u + \frac{v_z^2}{2}\right)\right] - Q - \rho_{mix} F_z v_z +$$

$$+ \frac{\partial}{\partial z} p v_z - \frac{\partial}{\partial z}(\tau_z \cdot v_z) + \frac{\partial}{\partial z} q_z'' = 0 \tag{63}$$

where: $\rho_{mix} = \alpha \rho_g + (1 - \alpha) \rho_l$. The solution of such equations requires the knowledge of the state equation.

If a turbulent flow has to be analyzed, in analogy to single-phase flows, the following relationships must be considered:

$$v_z = v_z' + \bar{v}_z; \quad \tau_z = \tau_z' + \bar{\tau}; \quad p = p' + \bar{p}; \quad u = u' + \bar{u} \tag{64}$$

where the variables with apices indicate the fluctuating values, while the overlined variables indicate the time-averaged values.

To solve conservation equations, in the case of a turbulent flow, it is necessary to use additional relationships (closure models), in analogy with the single-phase flows (see Section III).

2. *Zuber-Findlay Model (Two-Velocity Bidimensional Model)*

Zuber and Findlay have defined a local slip velocity, $\bar{v}_{gj}^{X}$, following the relationship:[11]

$$\overline{v_{gj}}^{X} = \overline{v_g}^{X} - j = (1 - r_g)(\overline{v_g}^{X} - \overline{v_l}^{X}) \tag{65}$$

where:

$$j = r_g \overline{v_g}^{X} + (1 - r_g) \overline{v_l}^{X} \tag{66}$$

where symbol $-X$ represents a weighted time-averaging (see Equation 53) and r_g is the fraction of residence time (see Equation 51) of the gaseous phase.

The following relationship may also be written:

$$r_g \overline{v_{gj}}^{X} = r_g \overline{v_g}^{X} - r_g j \tag{67}$$

Averaging in space Equation 67, the following is obtained:

$$\langle r_g \overline{v_{gj}}^{X} \rangle = \langle r_g \overline{v_g}^{X} \rangle - \langle r_g j \rangle \tag{68}$$

Owing to the fundamental relation in Equation 58, it may be written:

$$\frac{\langle r_g \overset{-X}{v_{gj}} \rangle}{\langle r_g \rangle} + \frac{\langle r_g \cdot j \rangle}{\langle r_g \rangle} \cdot \frac{\langle j \rangle}{\langle j \rangle} = \frac{\overline{\alpha \langle v_g \rangle}}{\langle r_g \rangle} \tag{69}$$

Since:

$$J_g = \overline{\alpha \langle v_g \rangle}; \quad \langle j \rangle = J; \quad \langle r_g \rangle = \bar{\alpha},$$

it is

$$\tilde{v}_{gj} + C_o J = J_g/\bar{\alpha} \tag{70}$$

where the sign $\sim$ means a weighting on r_g, from which:

$$\bar{\alpha} = J_g/(C_o J + \tilde{v}_{gj}) \tag{71}$$

where the "distribution parameter" C_o and the weighted slip velocity $\tilde{v}_{gj}$ are defined by:

$$C_o = \langle r_g j \rangle / (\langle r_g \rangle \langle j \rangle) \tag{72}$$

$$\tilde{v}_{gj} = \langle r_g \overset{-X}{v_{gj}} \rangle / \langle r_g \rangle \tag{73}$$

Some formulae for C_o and $\tilde{v}_{gj}$ in function of the flow regime, are reported in Chapter 3.

Equation 71 may also be written in terms of volumetric quality β:

$$\bar{\alpha} = \beta/(C_o + \tilde{v}_{gj}/J) \tag{74}$$

It may be noted from Equation 70 that in a flow where C_o and $\tilde{v}_{gj}$ are constant, the average velocity of the gaseous phase is a linear function of the superficial velocity J. With suitable values of parameters C_o and $\tilde{v}_{gj}$, other models proposed in the literature may be obtained.

3. Separated (Two-Phase) Flow Models

The separated flow models, analyzing separately the two phases and considering simplified phase-interaction mechanisms through empirical correlations, are characterized by generally limited validity ranges. Several models exist, differing substantially in the choice of the assumed empirical correlations. A relevant example of this category is proposed in Reference 12, used in the RELAP 5/mod. 1 computer program, with the following assumptions:

1. The flow is stationary and one-dimensional
2. In every section of the flow the phases are at the same pressure
3. The gaseous phase is in thermal equilibrium (at the saturation pressure)
4. The two phases may be in thermal and hydrodynamic disequilibrium

The following relationships (conservation equations) are proposed:

- Mass, liquid phase

$$\frac{1}{\Gamma_l}\frac{d\Gamma_l}{dz} = \frac{1}{\rho_l}\cdot\frac{d\rho_l}{\partial z} + \frac{1}{v_l}\frac{dv_l}{dz} - \frac{1}{1-\alpha}\frac{d\alpha}{dz} + \frac{1}{A}\frac{dA}{dz} \tag{75a}$$

- Mass, vapor phase

$$\frac{1}{\Gamma_g}\cdot\frac{d\Gamma_g}{dz} = \frac{1}{\rho_g}\cdot\frac{d\rho_g}{dz} + \frac{1}{v_g}\cdot\frac{dv_g}{dz} + \frac{1}{\alpha}\frac{d\alpha}{dz} + \frac{1}{A}\frac{dA}{dz} \tag{75b}$$

where:

$$\Gamma_l = \rho_l\, v_l\, A_l; \quad \Gamma_g = \rho_g\, v_g\, A_g$$

Since $d\Gamma_g/dz$ is the evaporation flow rate in the unit space, it may be written:

$$d\Gamma_g/dz = -d\Gamma_l/dz = \Gamma(dX/dz) \tag{76}$$

where X is the mixture quality.
In general, it is

$$\rho_l = \rho_l(i_l, p); \quad \rho_g = \rho_g(p)$$

The vapor phase is treated as a real gas, while the liquid-phase compressibility:

$$\frac{d\rho_l}{dz} = \left(\frac{d\rho_l}{di_l}\right)_p \frac{di_l}{dz} + \left(\frac{d\rho_l}{dp}\right)_{i_l} \frac{dp}{dz}$$

is neglected.

- Momentum, vapor phase

$$\alpha\,\rho_g\,A\,v_g\frac{dv_g}{dz} = -\alpha\,A\frac{dp}{dz} - F_{lg}\,A_{wg} - F\,A - \eta(v_g - v_l)\,\Gamma\frac{dX}{dz} + $$
$$- \rho\cdot g\,\alpha\,A\cos\theta \tag{77a}$$

- Momentum, liquid phase

$$(1-\alpha)\,\rho_l\,A\,v_l\frac{dv_l}{dz} = -(1-\alpha)\,A\frac{dp}{dz} + F_{lg}\,A - F_{wl}\,A +$$
$$-(1-\eta)\,(v_g - v_l)\,\Gamma\frac{dX}{dz} - \rho g(1-\alpha)\cos\theta \tag{77b}$$

where Θ is the channel inclination angle with respect to vertical; F_{lg} is the specific force (per unit volume) acting on the liquid phase in the direction of flow (acting on the vapor phase in the reverse sense); F_{wl} and F_{wg} are the specific forces (per unit volume) due to the flow and to the channel walls and acting, respectively, on the liquid and on the vapor phases (with the exception of the dispersed-flow regime, where it is $F_{wg} \simeq 0$); and η is the phase-change effect (in general, a value of 0.5 is assumed). An expression for term F_{lg} is given in Chapter 4, while F_{wl} and F_{wg} may be evaluated using classical methods in simple conditions.

- Energy

$$\Gamma\left[\lambda + \frac{1}{2}(v_g^2 - v_l^2)\right]\frac{dX}{dz} + \left(X\,\Gamma\,\frac{di_g}{dp}\right)\frac{dp}{dz} + (X\,v_g\,\Gamma)\frac{dv_g}{dz} +$$

$$+ [(1 - X)\,\Gamma\,v_l]\frac{dv_l}{dz} + [(1 - X)\,\Gamma]\frac{di_l}{dz} + \Gamma\,g\cos\theta = 0 \qquad (78)$$

Since for the gaseous phase the saturation pressure is assumed, it is

$$\Gamma[\lambda + 0.5(v_g^2 - v_l^2)]\,dX/dz = d\dot{Q}/dz \qquad (79)$$

where the second member represents the heat transfer rate per unit channel through the interface (see Chapter 6).

The two-phase nonequilibrium model (separated flow model) consists of Equations 75 to 79, which must be solved in six unknowns: α, X, v_g, v_l, p, and i_l. This is possible when the constitutive equations giving F_{lg}, F_{wl}, F_{wg}, and $d\,\dot{Q}/dz$, are specified.

REFERENCES

1. **Bird, R. B., Stewart, W. E., and Lightfoot, E. N.,** *Transport Phenomena,* John Wiley & Sons, New York, 1960.
2. **Schlichting, H.,** *Boundary Layer Theory,* 7th ed., McGraw-Hill, New York, 1979.
3. **Eckert, E. R. G. and Drake, R. M., Jr.,** *Analysis of Heat and Mass Transfer,* McGraw-Hill, New York, 1972.
4. **Yang, R. J. and Aung, W.,** Equations and Coefficients for Turbulent Modeling, NATO, Izmir, Turkey, 1984.
5. **Delhaye, J. M.,** Basic equations for two-phase flow modeling in *Two-Phase Flow and Heat Transfer in the Lower and Process Industries,* Bergles, A., et al., Eds., Hemisphere Publishing, New York, 1983.
6. **Banerjee, S.,** Conservation equations, and closure relationships and constitutive equations, in Short Course on Two-phase Flow and Heat Transfer, ETH-Bibliothek, Zurich, 1985, chap. 3 and 4.
7. **Sha, W. T., Chad, B. T., and Soo, S. L.,** Time Averaging of Volume-Averaged Conservation Equations of Multiphase Flow, AIChE Symp. Heat Transfer, Vol. 79, American Institute of Chemical Engineers, New York, 1983.
8. **Dobran, F.,** A two-phase fluid model based on the linearized constitutive equations, in Advances in Two-Phase Flow and Heat Transfer, NATO Adv. Res. Workshop, West Germany, 1982.
9. **Bankoff, S. G.,** A variable density single-fluid model for two-phase flow with particular reference to steam-water flow, *J. Heat Transfer,* 1960.
10. **Wallis, G. B.,** Some hydrodynamics aspects of two-phase flow and boiling, in International Developments in Heat Transfer, American Society of Mechanical Engineers, New York, 1963.
11. **Zuber, N. and Findlay, J. A.,** Average volumetric concentration in two-phase flow systems, *J. Heat Transfer,* 1965.
12. **Richter, H. J.,** Separated two-phase flow model: application to critical two-phase flow, *Int. J. Multiphase Flow,* 9(5), 1983.
13. **Vernier, P. M. and Delhaye, J. M.,** General two-phase flow equations applied to thermohydrodynamics of boiling water nuclear reactors, *Energie Primarie,* 4(1-2), 1968.

Chapter 3

FLOW REGIMES

I. INTRODUCTION

In two-phase flow fluidodynamic and thermal problems, the knowledge of the phase distribution inside the containment (channel, pipe, HX equipment) is of primary importance. Several typical flow situations have been identified (flow regimes), where different interactions between the two phases and between each phase and the containment wall may be recognized; this leads to different overall behaviors of the two-phase mixture. The most relevant parameters affected by the nature of the flow pattern (distribution of the phases in the flow or flow regime) are pressure drops and heat transfer coefficients (friction between fluid and wall varies greatly with the nature of the phase in contact with the wall; heat transfer greatly depends on the nature of the phase in contact with the wall, its velocity, etc.).

In this chapter, flow regimes typical of the most common applications are considered, together with methods for the evaluation of the void fraction in two-phase systems and of the conditions for flooding. These last two aspects are included owing to the characteristics of the evaluation methods presently used, based on phase-interaction analyses, in connection with the flow regime determination methods (Section II.A).

II. FLOW REGIMES

The importance of predicting the actual flow regime both in a piping system and in a heat exchange apparatus comes from the necessity of limiting the uncertainties at a design stage, especially as far as stability aspects or the possibility of unsafe operation conditions are concerned. Such aspects are connected to the relevant pressure drop variations that may follow a variation of flow regime (Chapter 4). In some cases, as in natural circulation systems design, the functionality of the system itself depends on the specific flow regime (in a natural circulation system, for instance, a separation of the two phases must be absolutely avoided).

A. Analytical Methods and Influence Parameters

The analysis of the different two-phase mixtures flow regimes initially was aimed at explaining the trends of heat transfer coefficients and of friction losses, with reference to a series of flow conditions of interest applications. In particular geometries and flow conditions, the necessity of a good knowledge of the local distribution of the phases was evident for reasons not concerning thermohydraulics, as in nuclear reactor design (effects of neutron moderation and absorption).

More recently, the necessity of dealing with complex mass and energy exchange phenomena between the phases in transient conditions and the tendency not to use specific energy models (for phase separation, critical velocity, flash phenomena, etc.), but to use general methodologies based on the application of the conservation equations for each phase, have imposed a detailed examination of two-phase mixture flow regimes. This is due to the necessity of a detailed definition both of thermal and mechanical interactions between the two phases and of the geometrical characterization of the relative contact surface.

The amount and the characteristics of each interaction depend on the spatial distribution of the two phases. As a consequence, a new approach and a new impulse to a deeper knowledge of flow regimes, requiring the overcoming of the traditional techniques of empirical characterization of the flow regimes, have become necessary. Present studies, therefore, aim more at the rationalization of results already available and at their phenomenological interpretation, than at the production of further experimental data. From an experimental point of view, however, flow regimes typical of complex geometries such as those of tube bundles, tees, etc., have still to be well-characterized. The repeatibility characteristics of the experiences have to be improved and a much wider and thorough analysis must be made on the influence of parameters such as the stabilization length and the duct diameter.

In the framework of this new and wider interpretation of the two-phase mixtures flow mechanisms, the themes of countercurrent flow of two phases and the limit conditions of flooding may also be treated. The kind of considerations developed in order to define the existence regions allows, at last, the determination of two extensively used parameters, the contact area between the two phases and (especially) the void fraction (see Section III). What concerns the parameters which influence flow regimes, the definition of the most suitable reference axes of the maps to be adopted for the representation of the existence zones of the various flow regimes, has represented a critical aspect, as it appears from the wide variety of coordinate couples assumed as reference for the representation of data in the literature. This is caused by the number and variety of parameters which influence the spatial disposition of the two phases. Among the most relevant parameters, with reference to a circular duct geometry, are the angle of slope with respect to the horizontal, the flow direction and the relative direction of the two phases, the average velocity of each phase, pressure and temperature, density and viscosity of each phase, surface tension duct diameter, and length of the duct. In complex geometries, instead of the diameter and length of the duct, other geometrical parameters typical of the geometrical configuration have to be considered, such as — if the flow is plane — the characteristic dimensions in the two directions, parallel and orthogonal, to the flow.

Furthermore, whatever geometrical configuration is considered, it is necessary to distinguish between adiabatic and diabatic conditions, owing to the importance (in presence of heat transfer) of the wettability and of the superficial finishing of the walls. The orientation of the channel with respect to the horizontal has an influence owing to the importance, in the various configurations, of the behavior of the phases with respect to gravity. Correspondingly, the direction of the flow (i.e., up or down) and the relative direction of the two phases are important because, as an effect of the buoyancy forces, the gaseous phase naturally tends to an upward motion with respect to the liquid phase. The phases' velocities and the magnitude of the slip are otherwise in connection to the mechanical interaction between the phases.

In countercurrent flow conditions, limit situations exist in which the mechanical interaction between the two phases causes the complete blockage of one of them (flooding, carry-under).

The thermodynamic properties of the mixture, such as density and viscosity of each phase, surface tension, and the operation pressure and temperature which influence them, define the importance of gravity, viscosity, and deformation actions. The channel diameter or other characteristic dimensions in a normal direction (to the flow) are important only if they interfere significantly with the dimensions of the dispersed phase (drops, bubbles) in the continuous matrix. Therefore, the channel diameter always has an influence in slug flows, while it affects the bubbly flow only if its value is very little. The length of the duct, or a characteristic length in the direction of the flow, has an influence only if it interferes with the length necessary to the stabilization of the flow. The influence of the length of straight portions of channels may be relevant in horizontal piping layouts, while in vertical channels it is generally limited. Alternatively, the number of tube rows in cross flows through tube bundles and the number and the geometry of spacing grids in parallel flow in tube bundles, show a relevant influence.

The general effect of interferences of mechanical components (valves, orifices, baffles, etc.) is, in flow regime analyses, the reduction of the existence regions for the "intermediate" flow regimes (slug, plug, intermittent).

The presence of a thermal flux in one-component two-phase mixtures causes a continuous transport of mass between the phases and, therefore, a continuous variation of the quality that makes the stabilization of a stationary condition more difficult or impossible. Also, in the situation of a two-phase mixture with two components, the presence of a thermal gradient generally has an influence, because its effects are normally different for the single phases (gas and liquid), with consequent variations of density and the gravity effect along the channel.

The numerous parameters of influence already mentioned are not generally considered singularly, but grouped in dimensionless groups. The selection of the representative dimensionless groups is a function of the prevailing forces acting on the system, which often change with the flow regime. In general, the presence of different forces affecting the flow contemporaneously makes the graphical representation of flow-regime maps difficult; as a consequence, only two adimensional groups are not generally sufficient to represent completely a flow-regime existence zone or the transition between two flow regimes. The most used characteristic numbers and parameters adopted to represent the flow regimes are Froude number, Weber number, Reynolds number, Martinelli parameter, superficial velocities, and mixture velocity.

B. Experimental Determination of Flow Regimes

The experimental determination of flow regimes has been performed in the past by means of direct observation (integral technique). More recently, another technique was adopted in conjunction with the preceding one, in order to obtain more complete information: that of the statistical, local fluctuations analysis, which considers one or some parameters (density, pressure near the wall, void fraction, etc.) and analyzes their fluctuations using statistical methods such as auto-correlation functions, cross-correlation functions, probability density functions, etc. It is based on the consideration that each parameter has different fluctuation modes in the various flow regimes. Typical results of statistical analyses are indicated in Figure 1,[1] with reference to vertical channels, and Figure 2, with reference to horizontal channels.[2]

It has been verified that in order to characterize a flow regime, the contemporary measurement of pressure in many points and of void fractions leads to good results, allowing a wide spectrum of possible function combinations to be analyzed both in time and space domains. Owing to the influence of vibrations and noise in the plants, the need of accurate standards and procedures in the acquisition of data exists.

C. Determination Theories of Flow Regimes

Two theories are described here due to their physical significance; since they are not examined in detail, an analysis of the original references is suggested.

1. Boundary Layer Technique

In order to provide a unique criterion for the determination of the existence region of a two-phase regime to be used in the prediction of thermal and mechanical interaction parameters between the phases and in the evaluation of the interphase surface area, an original approach is suggested in Reference 3, based on the extrapolation to two-phase flows of concepts developed in single-phase flow analyses (mixing length, Reynolds analogy, laminar sublayer, etc). As flow regime classification parameters, the kinematic flow characteristics in the laminar sublayer are selected. These flow characteristics, obtained in terms of velocity profile, apparent viscosity and apparent density, are used to characterize the flow regimes and are then directly employed to explain the trends of two-phase mixture friction factor, heat transfer coefficient, and void fraction.

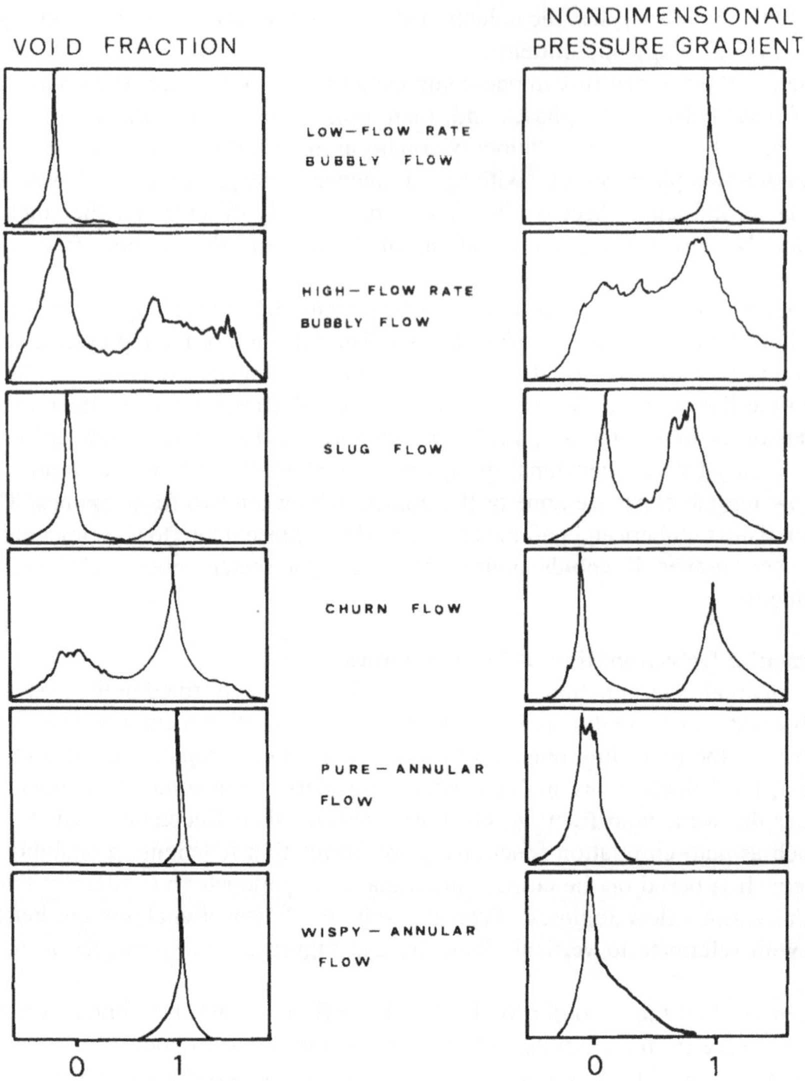

FIGURE 1. Void fraction and nondimensional pressure gradient (probability density functions).

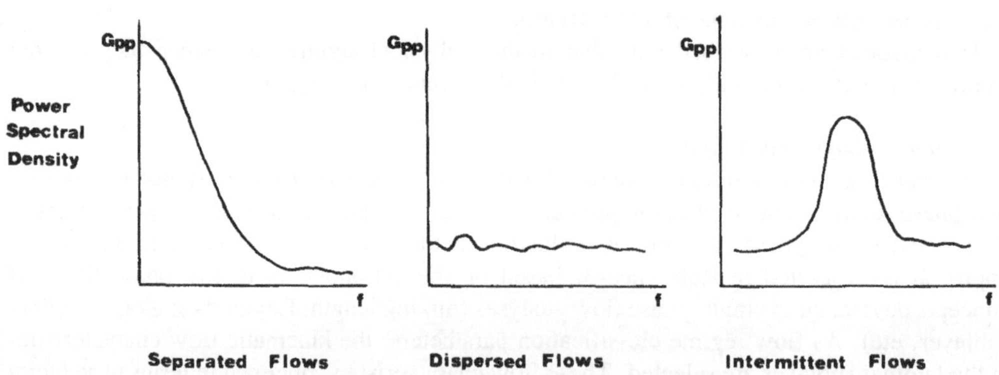

FIGURE 2. Relation between power spectral density function and flow type.

Table 1
SYNTHESIS OF BEATTIE FLOW-REGIME MODELS[3]

$$J = v_l(1 - \alpha) + v_g(\alpha)$$

$$j^+ = \frac{J}{\langle J \rangle}\sqrt{\frac{2}{f}}; \quad f = \frac{d(dp/dz)_l}{2\,\rho\langle J\rangle^2}; \quad Re = \frac{\rho\, d\,\langle J \rangle}{\mu}; \quad We = \frac{\rho\,\langle J\rangle^2\, d}{\sigma}$$

<J>: average value of J; δ^+: film thickness.

Sublayer type	Core region velocity profile
Viscosity-dependent flows (a, b, c, d)	$j^+ = a \ln y^+ + b; \quad y^+ = \frac{y}{R}\frac{Re}{2}\sqrt{\frac{f}{2}}$
Surface-tension-dependent flows (e)	$j^+ = a \ln y^+ + b; \quad y^+ = \frac{y}{R} We \frac{f}{Z}$
Film thickness-dependent flows (f)	$j^+ = a \ln\left[y^+\left(1 - \delta^+\left(1 - \frac{\mu_g}{\mu_l}\right)\right)^{-1}\right] + b;$ $y^+, \delta^+ = \frac{y, \delta}{R}\frac{Re}{Z}\sqrt{\frac{f}{2}}$

	Sublayer notation/type	Density definition	Viscosity definition
a.	Rigid surface bubbles	$\rho = \rho_f = \rho_l(1 - \beta) + \rho_g\beta$	$\mu = \mu_l(1 + 2.5\beta)$
b.	Nonrigid surface bubbles	$\rho = \rho_f = \rho_l(1 - \beta) + \rho_g\beta$	$\mu = \mu_l\left(1 + \frac{2.5\mu_g + \mu_l}{\mu_g + \mu_l}\beta\right)$
c.	Wavy gas/liquid interface	$\rho = \rho_f = \rho_l(1 - \beta) + \rho_g\beta$	$\mu = \mu_l(1 - \beta) + \mu_g\beta$
d.	Dry wall	$\rho = \rho_g$	$\mu = \mu_g$
e.	Attached wall bubbles	$\rho = \rho_f = \rho_l(1 - \beta) + \rho_g\beta$	N.A.
f.	Very thin film	$\rho = \rho_g$	$\mu = \mu_g$

The relationships proposed for the evaluation of mixture properties are indicated in Table 1, together with the hypothesized correspondence between the different relationships and the flow regimes.

2. *Prevailing Forces Technique*

An alternative analysis approach[4,6] systematically characterizes the possible and different existence regions for the two-phase regimes on the basis of the prevailing force(s) in each region. This technique allows a more effective and direct indication of the influence of the various parameters (listed in Section II.A) on the acting forces and, therefore, on the different flow regime existence regions. In usual applications, the acting forces are prevalently due to inertia, pressure, gravity, relative slip, and surface tension.

After analyzing the single forces and their ratios, it is possible to identify the prevailing adimensional parameters and, therefore, to determine the distribution of the phases (flow regime), the transition lines (regions) among flow regimes, or the possible instabilities between different adjacent flow regimes.

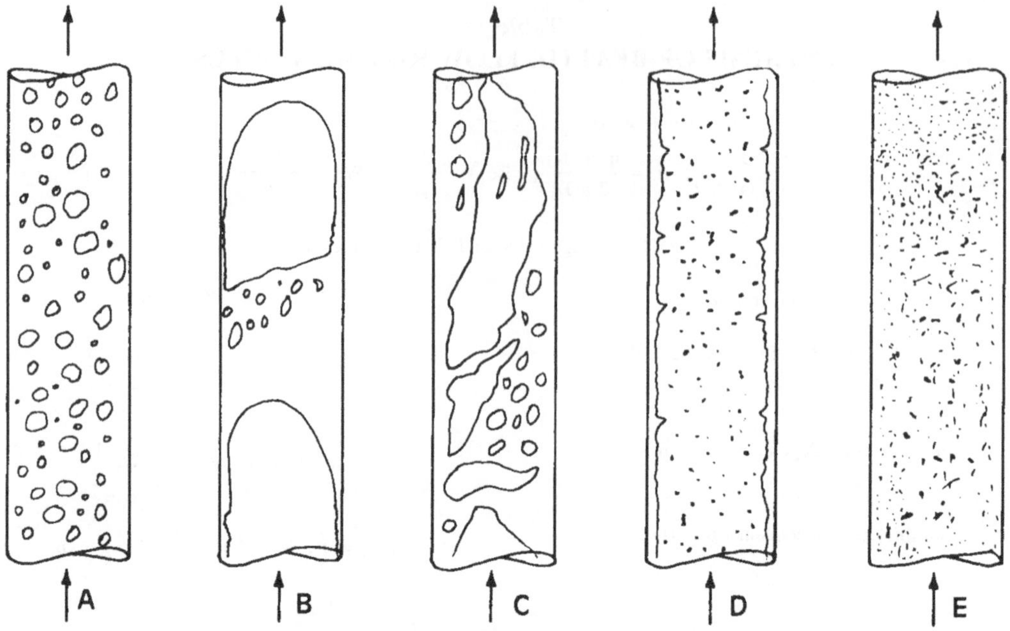

FIGURE 3. Flow regimes in vertical upflow. (A) Bubbly flow, (B) plug flow, (C) churn flow, (D) annular flow, (E) dispersed flow.

D. Flow Regime Maps in Adiabatic Conditions

Flow regime maps in adiabatic conditions are represented separately for vertical channels (upflow and downflow), horizontal channels and complex geometries (tube bundles, cross and parallel flow).

1. Vertical Channels

The terminology used in the definition of the various flow regimes is strongly dependent on the subjective interpretation of the visualized tests. A flow regime classification follows, reported only for practical purposes and valid only under a qualitative point of view (Figure 3).

1. Bubbly flow: This is characterized by a discontinuity in the gaseous phase, which is distributed in the liquid phase in the form of bubbles.
2. Plug flow: This occurs when the bubbles coalesce and reach the dimension of the transverse section of the channel. The vapor flow is separated from the liquid flow which contains small bubbles.
3. Churn flow: The plug flow may lead to instabilities; in this situation the shape of the plug is very irregular with the oscillatory phenomena of vorticous flow and with an unstable and oblong interface separation surface.
4. Annular flow: The liquid phase flows as a film close to the duct walls, with or without some bubbles. It is also present, in a dispersed form, in the gaseous phase flowing in the center of the duct.
5. Dispersed flow: The liquid phase is present as small drops dispersed in the gaseous phase, while the liquid film on the wall is no longer present. The dispersed flow is not generally obtained in adiabatic conditions, where an amount of liquid on the walls, although very small, is generally present.

The transition between the bubbly flow and the slug flow occurs owing to the bubble

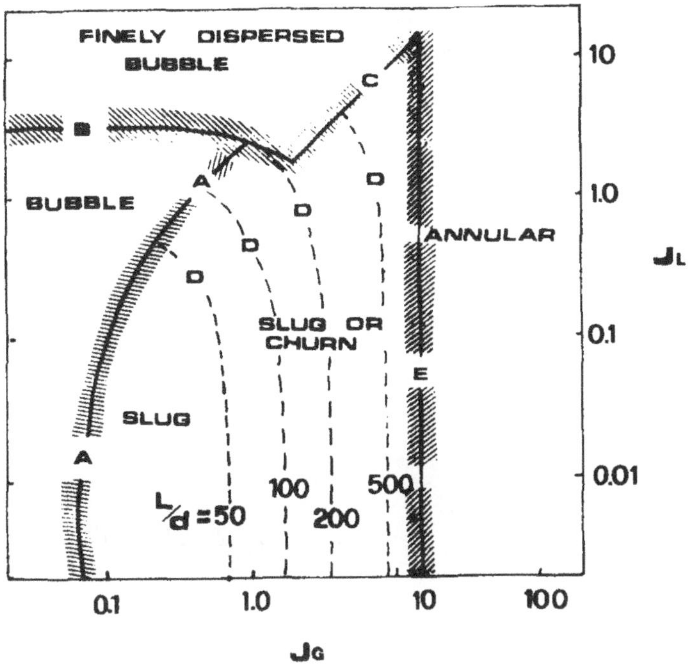

FIGURE 4. Vertical upflow regime map (d = 2.5 cm).

coalescence process, but is obstructed by the turbulent fluctuations which increase with the flow rate and lead to a greater probability of destruction of the bubbles. The equilibrium between the two processes defines, at each flow rate and pressure, the transition conditions. It may be assumed that with a limited dragging action of the fluid, the critical value of the void fraction at which the coalescence prevails is about 25%.

The transition to the annular flow is computed evaluating the minimum velocity of the gaseous phase, capable of drag the droplets with the maximum admissible dimensions in the specific conditions. In fact, with lower values of the gaseous phase velocity, the coalescence of drops may take place and liquid bridges appear, leading to the churn-slug-flow.

The upward vertical flow has been intensively studied both because of the simplicity of the geometrical condition and of the relevance in applications. The map reported in Figure 4 (see Reference 4; integrated with curve C) has particular importance from an applicative point of view. It is the result of one of the most recent and relevant studies under the auspices of the interpretation of regime transition mechanisms. The flow regimes are characterized as zones of prevalence of one of the forces acting on the system, and the transition "lines" are obtained from the equilibrium among the forces (see Section II.C.2). The transition conditions are reported in Table 2.

Checking the proposed correlation (Table 2) has been done using channels with different diameters (2.5 and 5 cm), with air-water mixtures in standard conditions. A particular feature of this map is the importance of the parameter L/d on the transition slug/churn, where L is the distance traversed by the mixture to the observing point. Another particularity is related to smaller diameters (less than 5 cm), where the bubbly zone, with low superficial velocities of the gaseous phase, becomes a slug zone. This map has been checked by many researchers, leading to satisfying indications with applicative conditions rather different from the original conditions.

The difference between the vertical upflow and downflow is caused by the different relative direction of the gravity force with respect to the inertial force. It leads to a differentiation

Table 2
TRANSITION CURVES OF FIGURE 4

Curve A (see note)

$$J_l = 3.0\, J_g - 1.15\left[\frac{(\rho_l - \rho_g)\, g\, \sigma}{\rho_l^2}\right]^{1/4}$$

Curve B

$$J_l + J_g = 4.0\left\{\frac{d^{0.429}\left(\dfrac{\sigma}{\rho_l}\right)^{0.089}}{\left(\dfrac{\mu_l}{\rho_l}\right)^{0.072}} \cdot \left[\frac{g(\rho_l - \rho_g)}{\rho_l}\right]^{0.446}\right\}$$

Curve C

$$J_l = J_g - 0.765\left[\frac{(\rho_l - \rho_g)\, g\sigma}{\rho_l^2}\right]^{1/4}$$

Curve D

$$\frac{L}{d} = 40.6\left(\frac{J_g + J_l}{\sqrt{g\, d}} + 0.22\right)$$

Curve E

$$J_g\, \rho_g^{1/2} = 3.1[(\rho_l - \rho_g)\, g\, \sigma]^{1/4}$$

Note: Curve A has no meaning for values of the diameter lower than a critical value, d_{cr}, expressed by: $\{\rho_l^2\, g\, d_{cr}^2/[(\rho_l - \rho_g)\sigma]\}^{1/4} = 4.36$.

of the characteristics of the various flow regimes and of the relative existence zones. The buoyancy force, in fact, always causes a tendency to an upflow of the gaseous phase; the importance of such an action is a function of the ratio between the amounts (masses) of the liquid and gaseous phases. More particularly (see Figure 5):

1. The structure of the bubbly flow is rather different from that in an upflow because the bubbles show a tendency to a concentration in the center of the channel.
2. The bubble agglomeration, increasing with the increase of the gaseous flow (with a constant liquid flow rate), leads to the formation of plugs, whose top is rounded and regular, while the basis is flat (irregular) and has bubble trains.
3. The bubbles may coalesce in oblong plugs (churn).
4. The churn bubbles may occupy the whole section and confine the liquid on the wall, originating in the annular regime, which may be subdivided (if the gaseous-phase flow increases) into falling film, bubbly falling film, and dispersed annular.

Such characteristics have been defined in Reference 5, on the basis of experimental data obtained with air and different liquids in a channel of 2.54 cm, at a pressure of 1.7 bar. The reference parameters for the proposed map (Figure 6) are

1. In abscissa: $x = \dfrac{Fr}{\sqrt{\lambda}}$ being $Fr = \dfrac{(J_g + J_l)^2}{g\, d}$ and $\lambda = \left(\dfrac{\mu_l}{\mu_g}\right)\left[\dfrac{\rho_l}{\rho_g}\left(\dfrac{\sigma}{\sigma_s}\right)^3\right]^{1/4}$

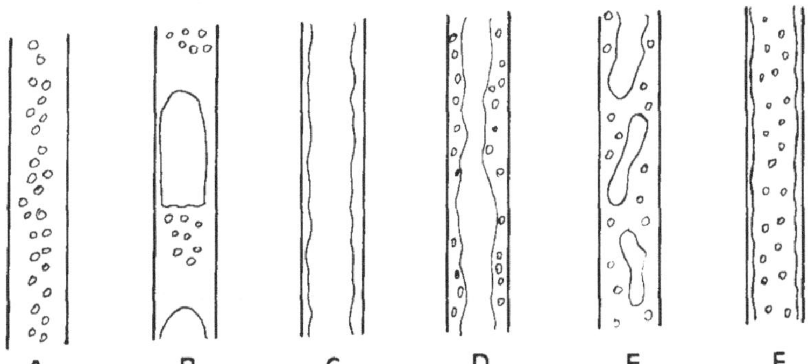

FIGURE 5. Flow regimes in vertical downflow. (A) Bubbly flow, (B) slug flow, (C) falling film flow, (D) bubbly falling film flow, (E) churn flow, (F) dispersed annular flow.

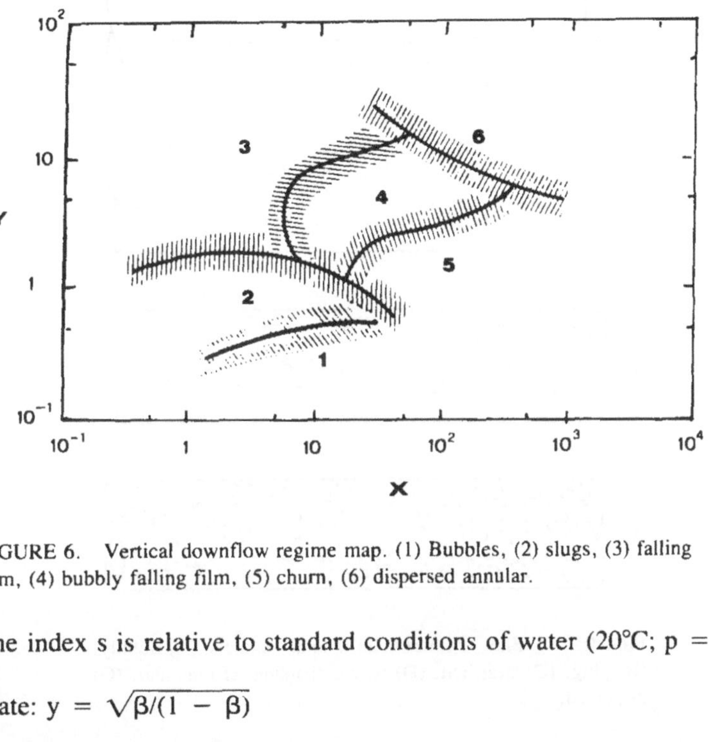

FIGURE 6. Vertical downflow regime map. (1) Bubbles, (2) slugs, (3) falling film, (4) bubbly falling film, (5) churn, (6) dispersed annular.

where the index s is relative to standard conditions of water (20°C; p = 1 bar).

2. In ordinate: $y = \sqrt{\beta/(1 - \beta)}$

where β is the volumetric vapor quality (see Chapter 1, Section IV.C).

2. Horizontal Channels

The main reason for the difference in two-phase flow regimes between vertical and horizontal channels is in the diversity of the effects of gravity force. In vertical channels, it acts in the same direction as the inertial and viscous forces; in the horizontal configuration, the gravity force acts in a direction normal for the axis of the channel. As a consequence of the different densities of the two phases, it may cause their stratification. In horizontal channels, all the possible flow conditions (even in the absence of a complete stratification) are affected by the orthogonality between the gravity force direction and that of the flow. This complicates the possible flow configurations of the mixture; with reference to Figure 7, the following flow regimes may be recognized:

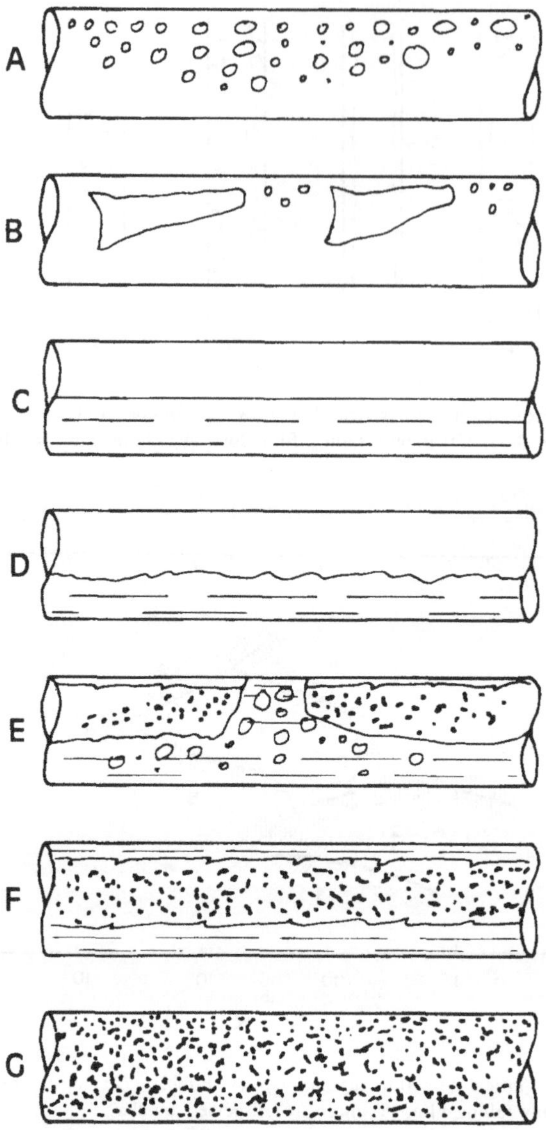

FIGURE 7. Flow regimes in horizontal channels. (A) Bubbly, (B) plug, (C) stratified, (D) wavy, (E) slug, (F) annular, (G) dispersed.

1. Bubbly flow: This corresponds to the analogous regime in vertical flow, with a greater bubble concentration in the higher part of the channel.
2. Plug flow: This is analogous to the corresponding vertical-channel regime, with the differences of a thicker liquid layer in the lower part of the channel and of an asymmetrical shape of the bubbles.
3. Stratified flow: This is characteristic of flows in horizontal and inclined channels. A sharp separation of the two phases occurs, with the liquid in the lower part and the vapor in the upper part of the duct.
4. Wavy flow: This represents a particular condition of stratified flow, where the interface surface between the two phases is rippled and irregular, owing to the presence of waves of limited dimensions.

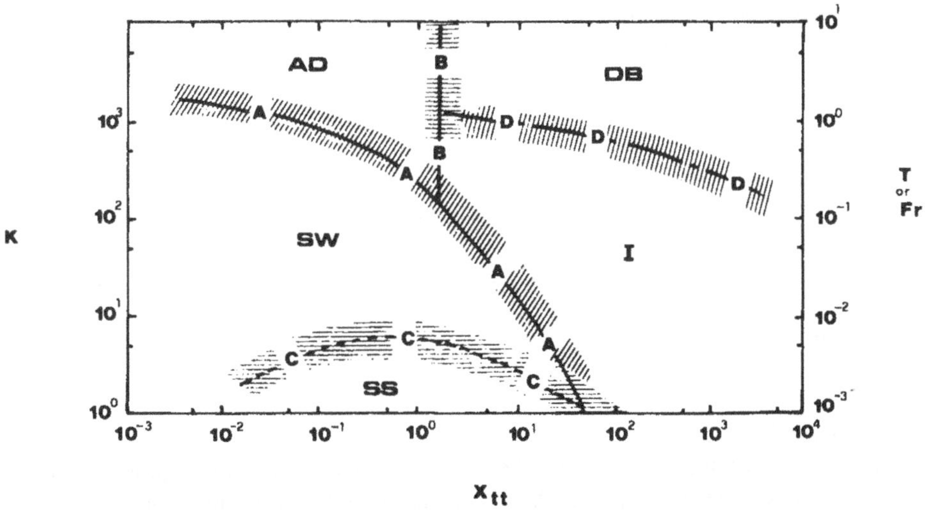

FIGURE 8. Horizontal flow regime map (A,B) Fr vs. X_{tt}, (C) K vs. X_{tt}, (D) T vs. X_{tt}). (AD: annular dispersed, DB: dispersed bubble, SW: stratified wavy, I: intermittent, SS: stratified smooth).

5. Slug flow: This represents a flow condition with separate phases (stratified), where the wave phenomenon on the interface is exalted. Waves reach the top of the duct and slugs of the liquid phase and the gaseous phase alternate.
6. Annular flow: This has analogous characteristics as the correpending flow in vertical channels, but the thickness of the liquid film is greater in the lower part of the channel.
7. Dispersed flow: It is similar to that of vertical channels.

An accurate analysis of regime transitions with flows in horizontal channels is reported in Reference 6.

The corresponding map (Figure 8) makes use of different coordinates for the different transitions, represented as functions of the Martinelli parameter (Chapter 1, Section V.A.16) and defined as:

$$\mathrm{Fr} = \sqrt{\frac{\rho_g}{(\rho_l - \rho_g)}} \cdot \frac{J_g}{\sqrt{d\ g \cos\theta}}; \quad K = \left(\frac{\rho_g\, J_g^2\, J_l\, \rho_l}{(\rho_l - \rho_g)\, g\, \mu_l \cos\theta}\right)^{1/2}$$

$$T = \left\{\left(\frac{dp}{dz}\right)_l / [(\rho_l - \rho_g)\, g \cos\theta]\right\}^{1/2}$$

where Fr is the generalized Froude number (Chapter 1, Section IV.A.11) (dp/dz) is calculated as if the liquid or the gaseous flow if the sole flow were in the duct; and Θ is the slope angle of the channel with respect to the horizontal.

Each of the transitions shown in Figure 8 has been interpreted in the light of force balances, using the methodology described in Section II.C.2. In particular, the transition between the wavy flow and the annular (or intermittent) flow is caused by the occurrence of instability conditions for the waves. In fact, the gaseous phase, in correspondence to a flow restriction (the top of the wave), accelerates and leads to a reduction of the local static pressure. This leads to an upflow motion of the liquid below, which is nevertheless obstacled by the gravity force. The acceleration of the gas increases with the wave dimension, but with it, the resistance due to gravity also increases. From the equilibrium of such forces, a criterion for the behavior of the waves may be obtained.

The transition between the stratified and the wavy regimes is coincident with the occurrence of waves on the interface surface, due to the effect of a drag induced by the faster gaseous phase.

The transition between the intermittent (plug flow) and the bubbly regime is identified as the condition in which, at high liquid and limited gaseous flows, the high turbulence in the liquid is able to prevail over the buoyancy forces which tend to maintain the gaseous phase concentrated at the top of the channel. As a consequence, the break of the gaseous pockets into small bubbles follows.

The proposed map[6] has been successfully compared with a great amount of experimental data. Further improvements are possible for the description of the different transitions; however, this map is one of the best results actually available, because it considers the influence of the channel diameter.

3. Inclined Channels

Systematic studies of the influence of the inclination angle on flow regimes have been performed by a few authors. Among the most significant results are those obtained in Reference 7, with inclined channels (downflow) in the range of 0 to 90°, and with a systematic extension of the interpretation of the transitions proposed by Taitel et al.[4,6] for vertical and horizontal channels.

With reference to flow regimes considered for horizontal channels, the greatest effect is a new existence region, that of the stratified flow with waves. In particular, as the slope angle increases with respect to the horizontal, the wavy flow region increases with respect to the corresponding situation in horizontal geometry. This effect is particularly important with angles between 0 and 10°, while the influence is limited between 10 and 70°; for higher values of the angle, the stratified flow disappears leaving an annular flow. The stratified flow without waves, on the contrary, is not noted with slope angles greater that 5°, as a consequence of the greater influence of the instabilities with respect to the horizontal configuration.

4. Tube Banks and Complex Geometries

The possibility of identification, representation, and repeatibility of two-phase flow regimes outside tube bundles is quite arduous owing to the greater chaoticity of the flow and to the presence of perturbation elements as the grids (in parallel flows) or the tubes themselves (in cross flows), which interfere with the tendency to a stabilized flow.

An analogous interference effect with the natural tendency to a stable asset of the two-phase flow occurs in other complex geometries whose behavior has been studied only recently (tees, bends, fittings, valves, etc.). The limited number of available experimental data and the great complexity of significant data acquisition make this matter the continuing object of analysis. Extrapolations (to complex geometries) of the relationships governing the transitions among flow regimes, obtained with reference to tubular ducts, have not yet been tried.

a. Tube Bundles, Cross Flow

Usually cross flow is of interest both for vertical and for horizontal flows. With reference to the test section of Figure 9,[8] the existence regions of the single regimes are shown in Figure 10, as function of parameters α^*, β^* defined as:

$$\alpha^* = J_l \frac{(\rho_l \, \mu_l)^{1/3}}{\sigma} \, [s^2/(m \, kg)]^{1/3}; \leftrightarrow \beta^* = J_g(\rho_g/\rho_l)^{0.5} \, [m/s]$$

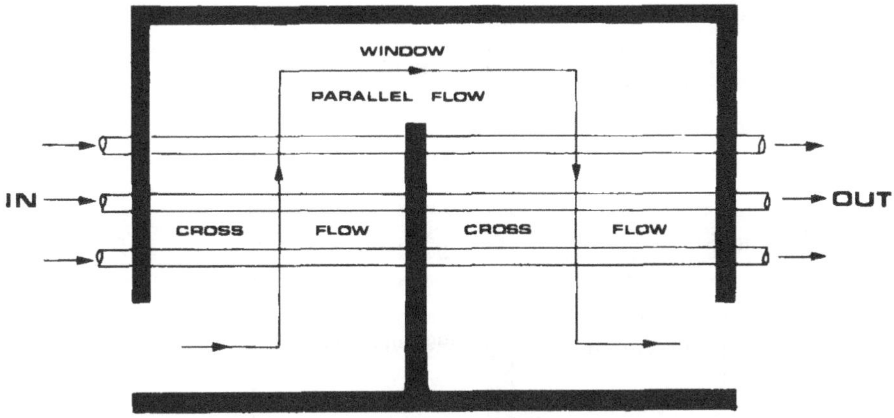

FIGURE 9. Experimental test section for ideal tube banks.

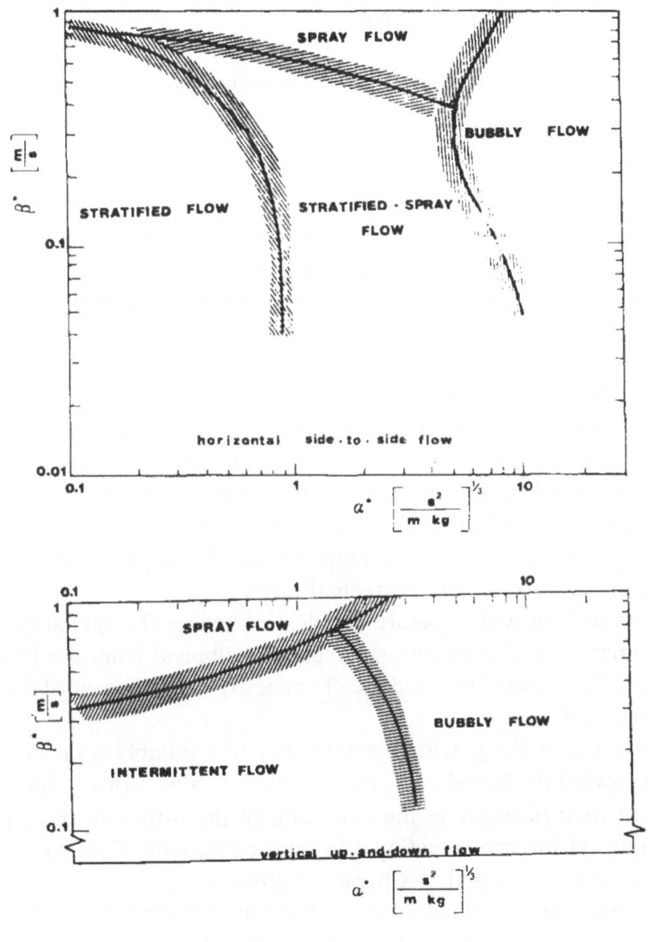

FIGURE 10. Cross-flow regime map.

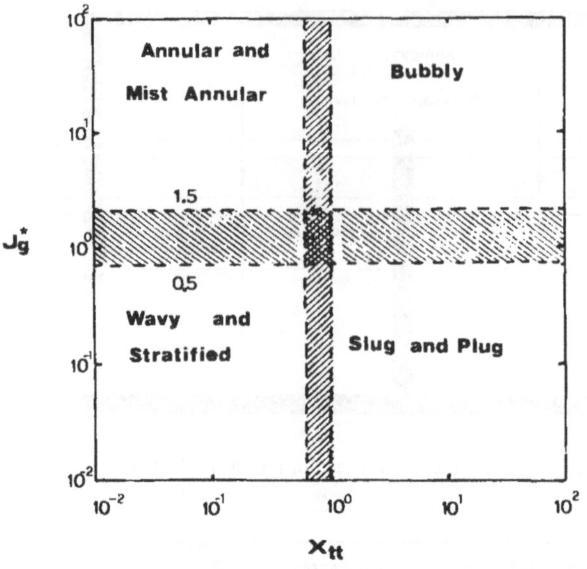

FIGURE 11. Horizontal tube-side diabatic (condensation) flow regime map.

b. Rod Bundles, Parallel Flow

An analysis of such configurations has been performed in Reference 10, with specific reference to nuclear power plant systems, reported here for completeness (analysis performed in diabatic conditions). Using a rod bundle of four heated rods in an unheated shroud, the contemporary presence (in the same transverse section) of different flow patterns characterized by a greater presence of liquid in the peripheral subchannels has been evidenced.

E. Flow Regime Maps in Diabatic Conditions

In each two-phase flow geometrical configuration previously examined, a contemporary heat transfer may exist, leading (in the relevant applicative situations of a single-component fluid) to vaporization or condensation. The influence of heat transfer on the existence regions of the flow regimes already identified in adiabatic conditions, has not yet been examined systematically for all the geometries. The most emphasis has been given to the study of condensation and boiling inside vertical and horizontal tubes.

The presence of nucleate boiling with separate bubbles influences the spatial distribution of phases through a systematic transfer of the vapor phase (bubbles) from the heated wall to the center of the channel. The shape itself and the characteristic dimensions of the bubbles are influenced by such an effect.

In other situations of nucleate boiling, with a coalescence of the bubbles on the wall and the presence of vapor jets toward the liquid core, the influence of a heat flux is more evident and may lead to consistent modifications in the extension of the different existence zones of the flow regimes recognized in corresponding adiabatic conditions. Evidence of bubble layers experimentally observed is reported in Chapter 9, Figure 11.

In the limit situation of subcooled thermal crisis on the wall (Chapter 10), a further two-phase flow regime exists, with a liquid core at the center of the channel and the vapor phase close to the wall (such a flow regime is seldom encountered in adiabatic flows; it may occur in critical flows of initially subcooled liquids).

Finally, the effects of boiling on the two-phase flow regimes extension and on transitions' characterization are a function of the intensity of the thermal flux transmitted by the wall and of the previously existing flow regime. Because of that, it is not possible to define general flow regime maps. It is possible, in some cases, to use the same maps as in adiabatic conditions, but great care must always be taken in extrapolations.

In condensation conditions, on the contrary, a systematic migration of vapor takes place from the flow core to the cooling walls.

The phenomenology is, then, quite different in the case of a flow inside or outside tubes, with evidence of a stronger coupling of the liquid and the vapor phases in flows inside tubes, in whatever flow regime. As in adiabatic conditions, gravity has a greater importance in horizontal flows, causing the tendency to stratified regimes.

The analytical approach to the study of the modes of flow in diabatic conditions has been oriented mainly to the problems of condensation inside tubes; in this phenomenon, in fact, the influence of the flow regimes on the heat transfer coefficient is stronger. The approach has been that of defining the influence of the diabaticity conditions on the existence regions recognized in adiabatic flows. Only the situation of horizontal channels with condensation is examined in this text.

In Reference 11, the validity of the flow regime map proposed by Taitel and Dukler[6] for adiabatic flows is also verified in diabatic conditions through a semitheoretical approach and by using a consistent set of experimental data. A simplification of such a map is also suggested, especially for its use in computer programs (Figure 11).

In the description of the map of Figure 11, the two fundamental flow conditions of gravity- and shear-controlled are distinguished, referring to the zones where the effect of gravity or the effect of interfacial shear stress between vapor and liquid prevails. The two zones are subdivided in two subzones, respectively, stratified-intermittent (gravity) and annular-bubbly (shear), to consider the influence of the vapor quality on each fundamental condition. The representation of diabatic flows using this approach is particularly significant if the parameters X_{tt} (Chapter 1, Section IV) and J_g^* are used, J_g^* being calculated as:

$$J_g^* = G\,X/\sqrt{d\,g\,\rho_v(\rho_l - \rho_v)} = [(Fa/Fr)\,(1/2\,f_v)]^{0.5}$$

where:

$$Fa/Fr = 2\,f_v\,G_V^2/[d\,g\,\rho_v(\rho_l - \rho_v)]$$ (ratio of vapor shear axial pressure gradient to gravity radial pressure gradient)

and f_v = friction factor with flow vapor only.

III. FLOODING

In two-phase downflows in vertical or inclined channels (tubes), particular conditions may occur, not completely described by the flow-regime maps. In particular, the condition of rising vapor, which blocks completely a descending liquid (flooding), may occur. This condition is preceded, with lower velocities of the vapor phase, by the phenomena of dragging liquid drops in a vapor upflow (carry-over), and with even lower values of the vapor velocity (flow) and greater descending liquid mass flow rate, by the phenomena of dragging the vapor downflow (carry-under) in the liquid flow.

Carry-under and carry-over processes are of particular interest in the study and design of moisture separator equipment, while flooding is a phenomenon of interest mainly in light water nuclear reactor safety. The flooding phenomenon develops, in annular flow conditions, through the evolution of the superficial characteristics of the descending liquid layer. As the vapor velocity increases, the following configurations may be obtained: smooth film,

film with two-dimensional, three-dimensional, stationary, and moving waves. The determination of the transition conditions among such flow modes is necessary to characterize rigorously the conditions for flooding.

Another characteristic aspect of the occurrence of flooding is the sudden increase of the pressure losses that, verified by many researchers, corresponds to the appearance of big droplets transported in the vapor stream. This phenomenon also is interesting in defining the conditions for flooding. The characteristic aspects are pointed out in Figure 12, as the result of a wide experimental study performed in a rectangular geometry (Figure 13).[17]

The proposal of models for the phenomenon of flooding was originated by semiempirical studies aimed at phenomenological comprehension. The results of such a phase of analysis may be synthesized in a number of correlations semiempirical in nature, among which the following[14] is applicable when gravity forces are much more relevant than the viscous ones:

$$J_g^{*0.5} + m\, J_l^{*0.5} = C \quad (1)$$

where $m = 0.8$ to 1 and $C = 0.7$ to 1. J_g^* is the Wallis parameter:

$$J_g^* = J_g\, \rho_g^{0.5}/[g\, d(\rho_l - \rho_g)]^{0.5}; \leftrightarrow J_l^* = J_l\, \rho_l^{0.5}/[g\, d(\rho_l - \rho_g)]^{0.5}$$

After a phase in which semiempirical models were proposed, another followed, in which the phenomenon was studied with the aim of analytically representing its peculiar aspects.

A trial has been completed to explain the instability of the superficial waves of the film in terms of a force balance, or to explain such a situation as the passage from annular or stratified to slug flow conditions. Alternatively, the interaction between the two phases has been interpreted in terms of the transfer of momentum through a rough and pervious wall, determining the flooding phenomenon as the limit condition of existence of a countercurrent flow.

A review of flooding models[12] classifies them as follows: (1) models based on the envelope of "existence conditions" (semiempirical correlations, as in Equation 1); (2) models based on the stability of superficial waves; (3) models based on the static equilibrium of forces on the film; and (4) models based on the interpretation of flooding as a transition slug-annular or slug-stratified.

In the limits of the present text, mention is made only of the annular-slug transition approach. This approach, oriented in such a way as to recognize flooding in terms of transition between annular and slug flow, describes the condition in which a wave, formed on the liquid film, may become unstable and increase indefinitely until forming a slug. This criterion allows the use of models already developed for the description of the two-phase flow regimes (maps) and for horizontal or inclined channels, considering adequate the vertical countercurrent flow situation. Following Reference 6, the condition of flooding is described by:

$$v_g - v_l > C\{(\rho_l - \rho_g)\, g \cos\theta\, A_g/[\rho_g(d\, A_l/d\, h_l)]\}^{0.5} \quad (2)$$

where h_l is the thickness of the liquid layer; A_l and A_g are the liquid and gas areas; C may be evaluated as $C = 1 - h_l/d$; and $(d\, A_l/d\, h_l)$ is the derivative notation.

In the current research on flooding, particular effort is dedicated to analyzing flooding conditions with heat transfer, namely in condensation situations.

IV. VOID FRACTION

A parameter often used to macroscopically characterize the relative importance of the two phases is the void fraction, defined (in one-dimensional flows) as the ratio between the

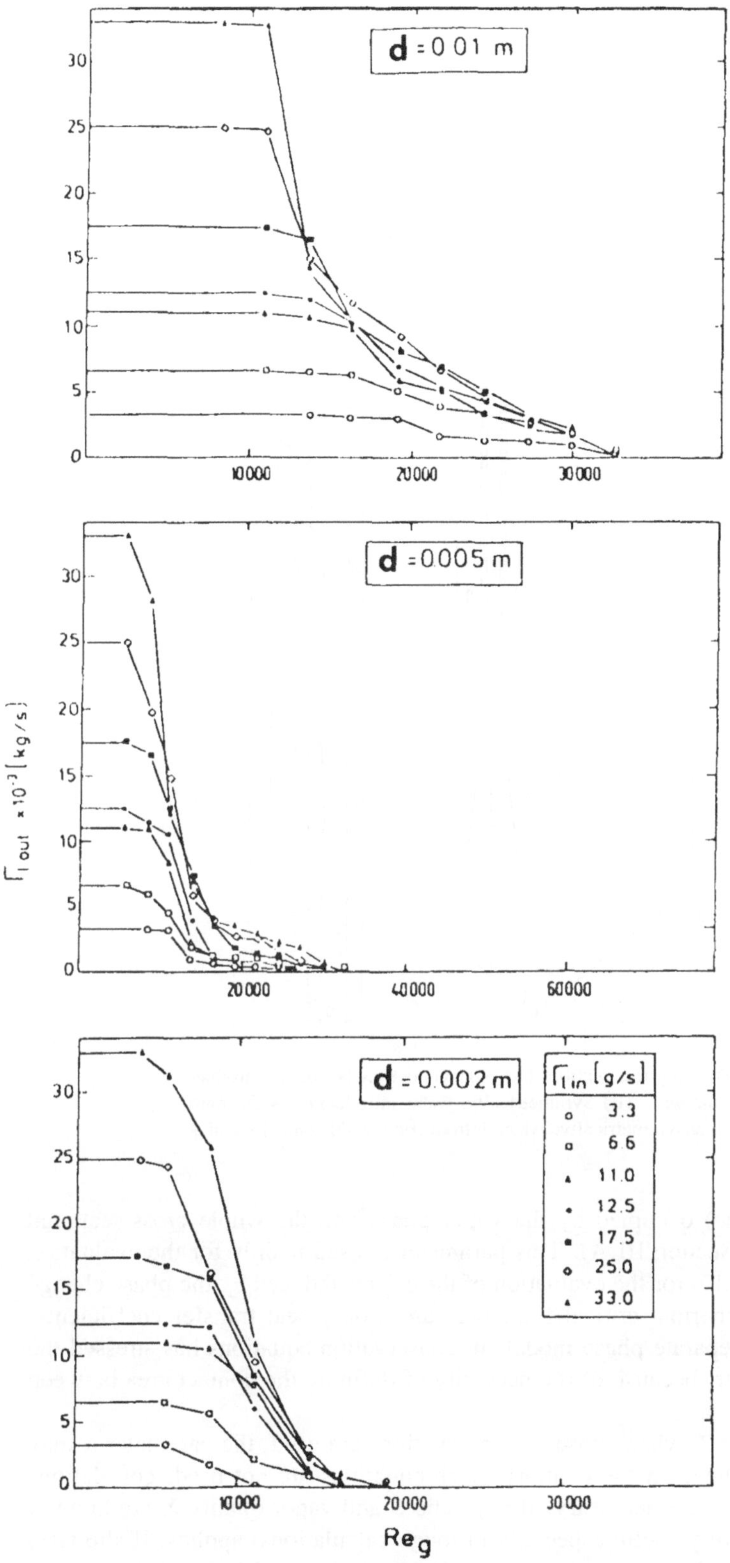

FIGURE 12. Outlet liquid mass flow rate vs. gas mass flow rate.

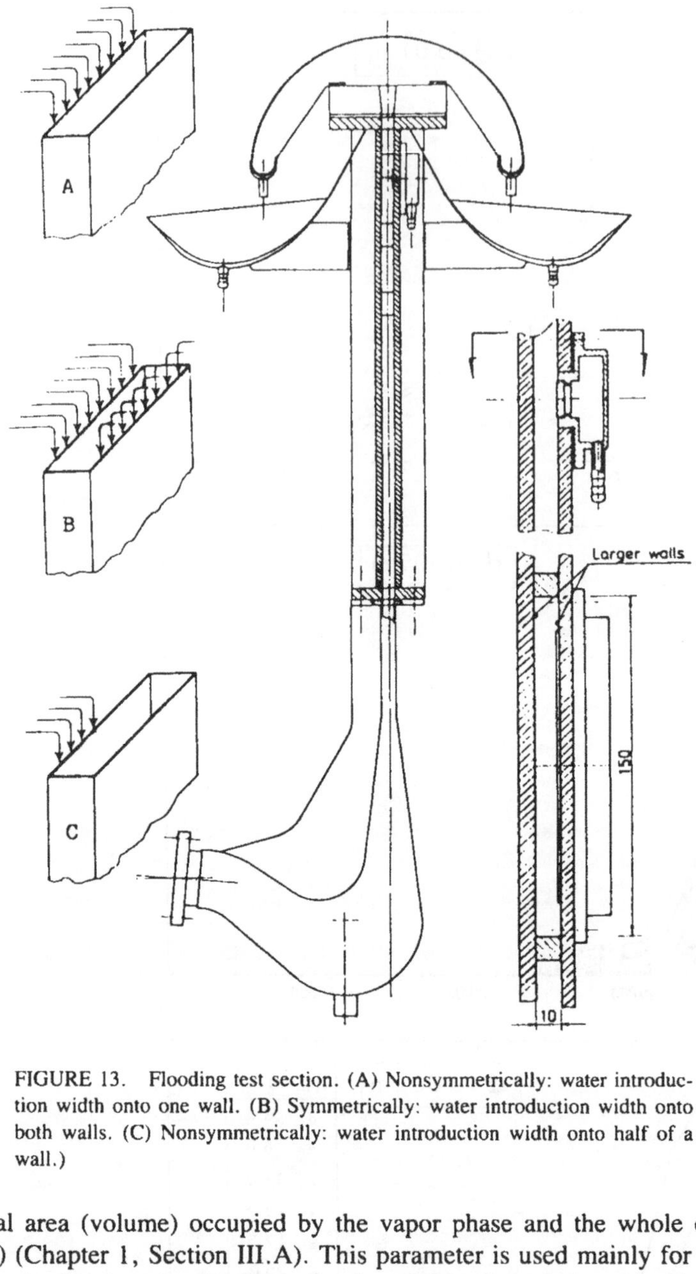

FIGURE 13. Flooding test section. (A) Nonsymmetrically: water introduction width onto one wall. (B) Symmetrically: water introduction width onto both walls. (C) Nonsymmetrically: water introduction width onto half of a wall.)

cross-sectional area (volume) occupied by the vapor phase and the whole cross-sectional area (volume) (Chapter 1, Section III.A). This parameter is used mainly for the evaluation of the mixture density, but also for the evaluation of the effects induced by the phase change in parameters such as the thermal critical flux, pressure drops, heat transfer coefficients. More recently, the use of separate-phase models in conservation equations has stressed the importance of this parameter, because of the necessity of defining the contact area between the phases.

As pointed out in Chapter 2, when conservation equations are used, the parameter α may be obtained from them. When, on the contrary, such equations are not used, correlations for the direct evaluation of α are necessary. If slip ratio S and vapor quality X are known, Equation 11 of Chapter 1 (very useful especially in rough calculations) applies. If slip ratio S is not known (X may be easily evaluated), use must be made of one of the several empirical correlations proposed in the literature for the evaluation of this important parameter (their validity is limited to the experimental conditions which have defined them). As an example, an experimental correlation with a rather wide range of applicability and of interest also in diabatic conditions, is the following,[15] obtained making use of a thermal-neutron scattering technique. This correlation relates void fraction to quality, mass velocity, and the ratio of densities:

$$R \cdot X = [\alpha^2(Y^{1/3} - 1) + \alpha]/[Y - \alpha(Y - Y^{1/3})] \quad (3)$$

where:

$$R = 1 + 0.306(1 - X)^{2.105} \quad (4)$$

and

$$Y = 0.027(\rho_l/\rho_g)\ G^{0.525} \quad (5)$$

Figure 14 shows a comparison with experimental data, with reference to a range of pressure between 50 and 70 bar and to a range of mass velocity between 150 and 300 kg/(m^{2}sec) (quality ranging between 0.1 and 1).

In Chapter 2, Section IV.E.2, another model (slip model) has been proposed for the evaluation of the void fraction, allowing a particularly accurate representation of such a parameter. It is suitable to application in the field of separate-phase flows, where the evaluation of the actions transmitted at the interface is particularly important. Such a model is based on the definitions of two parameters, $\tilde{v}_{gj}$ (relative flow velocity of the vapor phase with respect to the mixture) and C_o (distribution-phase parameter), already defined in Chapter 2, Section IV.E.2.

As a consequence of the definition of $\tilde{v}_{gj}$ and C_o, it is

$$\alpha = X/\{C_o[X + (1 - X)\ \rho_g/\rho_l] + \rho_g\ \tilde{v}_{gj}/G\} \quad (6)$$

The parameter C_o substantially depends on the kind of flow regime, while $\tilde{v}_{gj}$ also depends on the thermodynamic and transport properties of the phases. The expression of C_o and $\tilde{v}_{gj}$ is discussed by many authors, and for some geometries they have not yet been defined.

According to Reference 16, it is $C_o = 1$, for a uniform distribution of the bubbles or of the drops; $C_o < 1$, for an incipient boiling; and $C_o > 1$, for all other flow conditions (C_o is always ≤ 1.5).

Parameter $\tilde{v}_{gj}$ may be evaluated as follows:

$$\tilde{v}_{gj} = 0.35[g\ d(\rho_l - \rho_g)/\rho_l]^{0.5}$$

for plug flows and:

$$\tilde{v}_{gj} = 1.53[g\ \sigma(\rho_l - \rho_g)/\rho_l^2]^{0.25}$$

for bubbly flow.

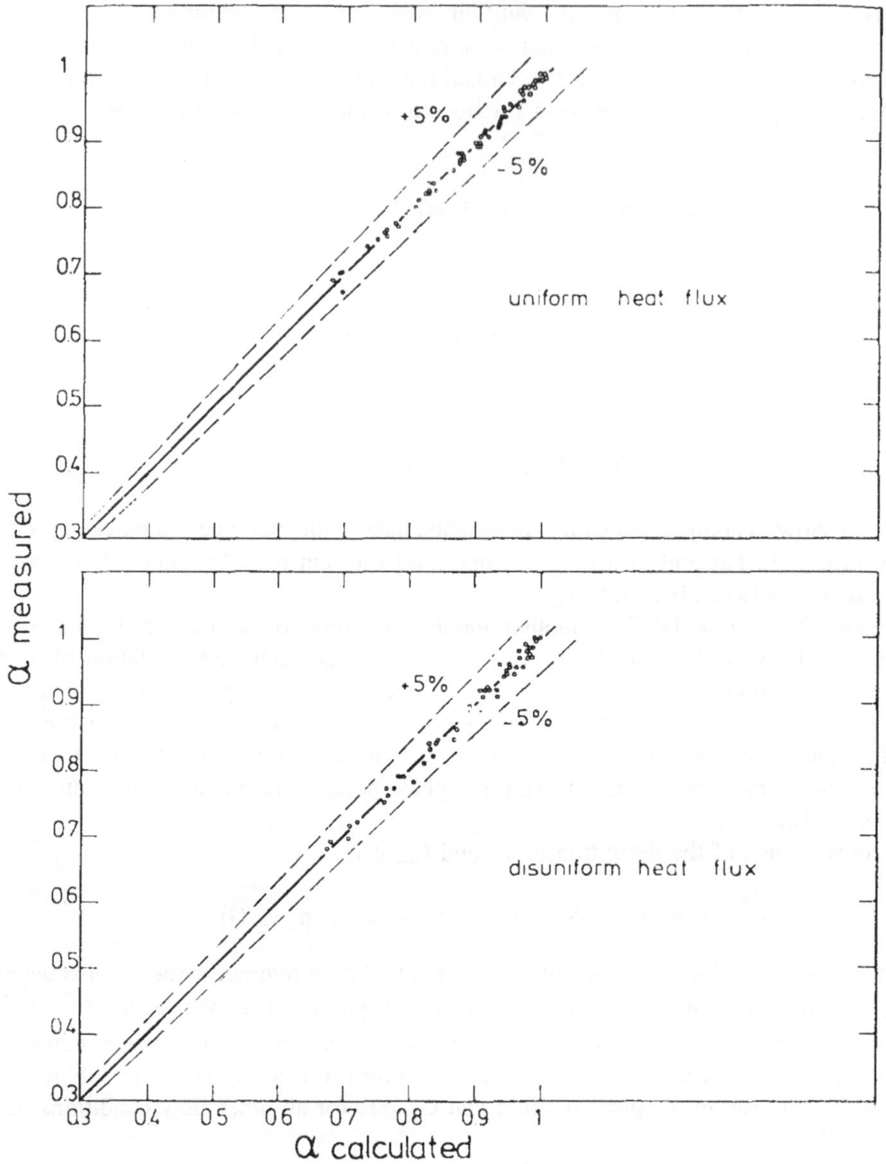

FIGURE 14. Comparison between Equation 5 and experimental data.

REFERENCES

1. **Annunziato, M. and Girardi, G.,** Flow pattern identification by pressure drop and void fraction fluctuations analysis in vertical two-phase flow, paper presented at European Two-Phase Flow Group Meeting, Rome, June 19, 1984.
2. **Hubbard, M. G. and Dukler, A. E.,** The characterization of the flow regimes for horizontal two-phase flow: statistical analysis of the wall pressure fluctuation, *Proc. of the 1966 Heat Transfer and Fluid Mechanics Institute*, Saad and Miller, Eds., Stanford University Press, Stanford, Calif., 1966.
3. **Beattie, D. R. M.,** An Extension of Single Phase Flow Turbulent Pipe Flow Concepts to Two-Phase Flow, Ph.D. thesis, University of New South Wales, Sydney, Australia, 1983.

4. **Taitel, Y., Barnea, D., and Dukler, A. E.,** Modelling flow pattern transitions for steady upward gas-liquid flow in vertical tubes, *AIChE J.*, 26(3), 1980.
5. **Oshimowo, T. and Charles, M. E.,** Vertical two-phase flow. I. Flow pattern correlation, *Can. J. Chem. Eng.*, 52, 1974.
6. **Taitel, Y. and Dukler, A. E.,** A model for predicting flow regime transitions in horizontal and near horizontal gas-liquid flow, *AIChE J.*, 22, 1976.
7. **Barnea, O., Shoham, O., and Taitel, Y.,** Flow pattern transition for downward inclined two-phase flow, *Chem. Eng. Sci.*, 37(5), 1982.
8. **Chisholm, D. and Grant, I. D. R.,** Two-phase flow on the shell-side of a segmentally baffled shell and tube heat exchanger, *J. Heat Transfer*, 101, February 1979.
9. **Bell, K. J., Taborek, J., and Fenoglio, F.,** Interpretation of horizontal channels in tube condensation heat transfer correlations with a two-phase flow regime map, *Chem. Eng. Prog. Symp. Ser.*, 66, 1970.
10. **Bergles, A. E.,** Two-phase flow structure observations for high pressure water in a rod bundle — in two-phase flow and transfer in rod bundle, *ASME J.*, 47, 1969.
11. **Palen, J. W., Breber, G., and Taborek, J.,** Prediction of flow regimes in horizontal tube — side condensation, *Heat Transfer Eng.*, 1(2), 1979.
12. **Bankoff, S. G. and Lee, S. C.,** A critical review of the flooding literature, NUREG/CR - 3060 - July 1983.
13. **Tien, C. L., Chung, K. S., and Lin, C. P.,** Flooding in two-phase countercurrent flows. I. Analytical modelling, *Physiochem. Hydrodynam.*, 1, 1980.
14. **Wallis, G. B.,** *One-Dimensional Two-Phase Flow*, McGraw Hill, New York, 1969.
15. **Cumo, M., Frazzoli, F. V., Palazzi, G., and Urbani, G.,** Fullscale tests on axial profile heat flux influence on the critical quality in PWR steam generators, CNEN, RT/ING, (80)5, 19.
16. **Zuber, N. and Findlay, J. A.,** Average volumetric concentration in two-phase flow systems, *J. Heat Transfer*, November 1965.
17. **Cumo, M., Farello, G. E., and Furrer, M.,** Flooding Conditions in Falling Film Rewetting, ENEA R.T.I., NEBH-1TS4B-84009, February 1984.

Chapter 4

PRESSURE DROPS

I. INTRODUCTION

Pressure drops represent one of the most studied and relevant themes of thermohydraulics. Their evaluation is necessary in the design of all fluid systems and is required in functional verifications and in safety analyses of a great variety of components and systems.

In all fluid systems, pressure drops are responsible for the required pumping power; in parallel channels, they influence the flow distribution among the channels; in two-phase flows, an accurate evaluation of pressure drops is particularly important for the possibility of unstable conditions.

The wide number of geometries and situations of interest in applications makes the review of the proposed models and semiempirical correlations quite hard. The level itself of the knowledge is not uniform: while for single-phase flows the correlations and methods are rather satisfactory, in two-phase flows, where the phases interact both mechanically and thermally, other information is necessary in order to develop satisfactory overall models.

To this aspect and to a deeper analysis of complex geometries (i.e., steam generators) research is presently mainly devoted.

II. SINGLE PHASE

The treatment of single-phase flow pressure drops refers both to liquid and gas or vapor flows. The aspects requiring further experimental and theoretical analyses mainly concern special geometries.

A. Flow Inside Tubes

Single-phase flow inside tubes represents the most common fluidodynamic situation, involving a great number of heat exchange apparatuses and the maximum number of applications of transport of fluids.

The pressure drops of a single-phase fluid flowing in a duct may be expressed, in its most general form, by the following relationship:

$$\Delta p_{tot} = \Delta p_f + \Delta p_{obs} + \Delta p_{acc} + \Delta p_{elev} \tag{1}$$

Δp_f represents the pressure losses due to fluid/wall contact friction, which are distributed along the length of the duct. Their evaluation is possible using Equation 2, semiempirical in nature and obtained through the application of momentum conservation (see Chapter 2):

$$\Delta p_f = 4\ f(L/d_i)\ \rho\ v^2/2 \tag{2}$$

In Equation 2, f is the Fanning friction factor (Section II.A.1), d_i is the internal diameter of the tube when the geometry is circular; with other geometries, d_i is the equivalent hydraulic diameter, expressed by:

$$d_i = d_w = 4\ A/P_w \tag{3}$$

where A is the transverse flow area and P_w the wetted perimeter (Equation 3 applies correctly in complex geometries, only with a uniform distribution of solid bodies in the transverse flow area; also see the footnote in Chapter 8, Section III.B).

Δp_{obs} represents the irreversible pressure drops due to obstacles and variations in the geometry or in the configuration of the duct. In analogy to the term Δp_f, for the term Δp_{obs} the following expression may be used:

$$\Delta p_{obs} = \xi \, \rho v^2/2 \tag{4}$$

ξ is the concentrated losses factor, to be evaluated as indicated in Section II.A.2.

Δp_{acc} represents the pressure variation due to a variation in the kinetic energy of the fluid. In a single-phase flow, such variation is generally important only for gases:

$$\Delta p_{acc} = \Gamma^2 \int_1^2 (1/A) \, d(1/\rho \, A) \tag{5}$$

where index 1 represents an upstream section and index 2 a downstream section with respect to the kinetic energy variation, i.e., the two sections where Δp is measured.*

Δp_{elev} represents the variation of pressure due to the variation in the potential energy of the fluid (in a gravitational field):

$$\Delta p_{elev} = \rho \, g \, \Delta z \tag{6}$$

where g is the gravity acceleration and Δz the variation of height between the two considered sections.

1. Continuous Friction

Friction factor f of Equation 2 is the so-called "Fanning factor", dependent on the characteristics of the flowing fluid (through the Reynolds number) and — in turbulent flow — also on the geometrical characteristics of the duct (through the relative rugosity, defined as the ratio between the average value of the superficial rugosity and the duct diameter). The dependence of the friction factor f on the Reynolds number may be obtained theoretically only in the situation of a laminar flow; in turbulent flows, it is determined through experimental analyses. For commercial tubes and adiabatic flows, the relationships for the evaluation of f are[1]

$$f = 16/Re \qquad \text{(laminar flow)} \tag{7a}$$

$$1/\sqrt{f} = -4 \log_{10}[\epsilon/(3.7 \, d_i) + 1.255/(Re\sqrt{f})] \tag{7b}$$

(turbulent flow)

The overall trend of f is shown in Figure 1 where, for smooth tubes, the curve for turbulent flow may be approximated by the following relationship[1]:

$$f = 0.046 \, Re^{-0.2}$$

$$5000 \leqslant Re \leqslant 200{,}000 \tag{8a}$$

$$1/\sqrt{f} = 4 \log_{10}(Re\sqrt{f}) - 0.4$$

$$Re > 200{,}000 \tag{8b}$$

* If $\rho \simeq$ const., $\Delta P_{acc} = \bar{\rho}(v_2^2 - v_1^2)/2$.

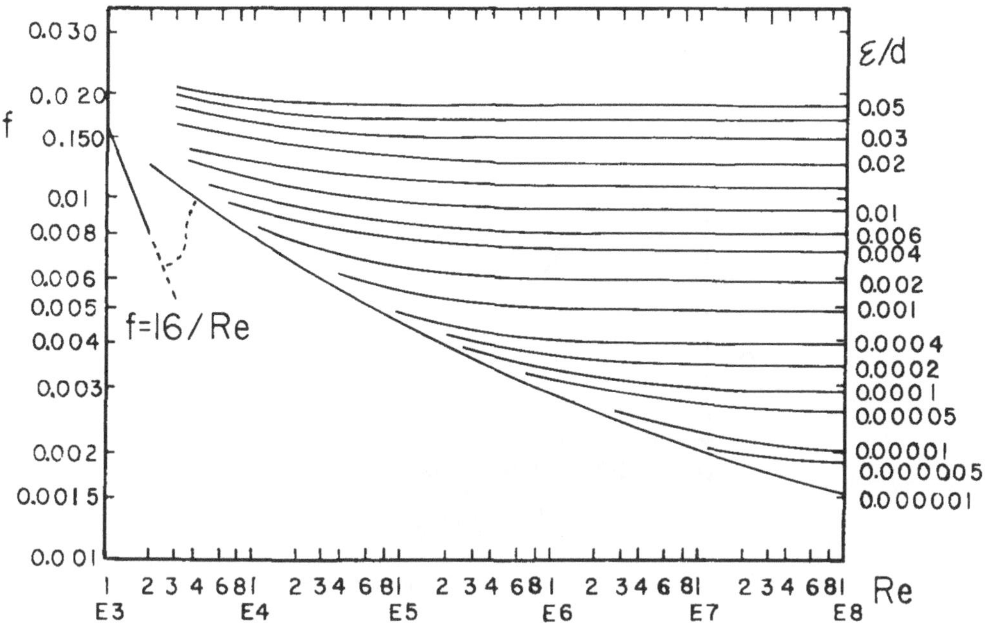

FIGURE 1. Fanning friction factor.

In diabatic conditions, the influence of the thermal gradient near the wall may be taken into account modifying the isothermal friction factor f_{iso} (evaluated using Equations 7 or 8) through the following:

$$f/f_{iso} = (Pr_w/Pr_b)^{a_1} \simeq (\mu_w/\mu_b)^{a_1} \tag{9}$$

where f is the nonisothermal friction factor; subscripts w and b indicate the evaluation of physical properties, respectively, at the wall and at the fluid-bulk temperature; and exponent a_1 is[2]

1. $a_1 = 0.23$ in laminar flow and $a_1 = 0.11$ in turbulent flow, in the case of cooling of the fluid
2. $a_1 = 0.38$ in laminar flow and $a_1 = 0.17$ in turbulent flow, in the case of heating of the fluid

2. Geometry Variations and Obstacles

A systematic treatment of pressure drops in the presence of obstacles and of complex geometries may be found in Reference 3. In the present text, only the most common situations are examined.

Nonreversible pressure drops (pressure losses) due to abrupt variations in the geometry of a flow channel are due to local fluid volumes with a high level of vorticity, which are a function of the typology of the geometry variation and of the turbulence of the upstream fluid.

If an abrupt variation of area occurs in a duct (from A_1 to A_2), the dependence of the concentrated loss factor ξ (Equation 4) on the Reynolds number and on the ratio between the upstream and downstream areas of the geometry variation is

$$\xi = \xi(Re, A_1/A_2) \tag{10}$$

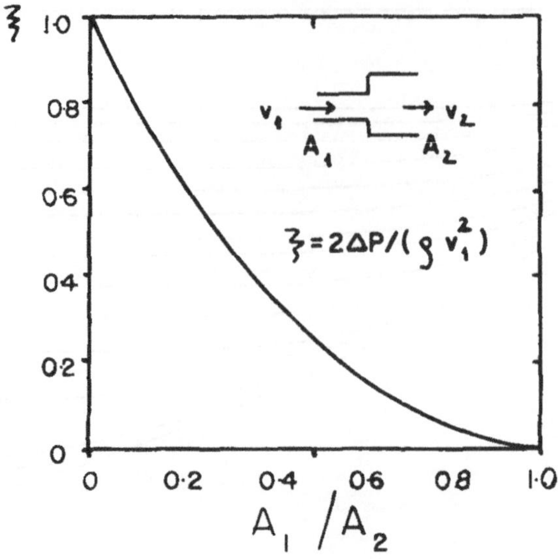

FIGURE 2. Concentrated loss coefficient (sudden enlargement).

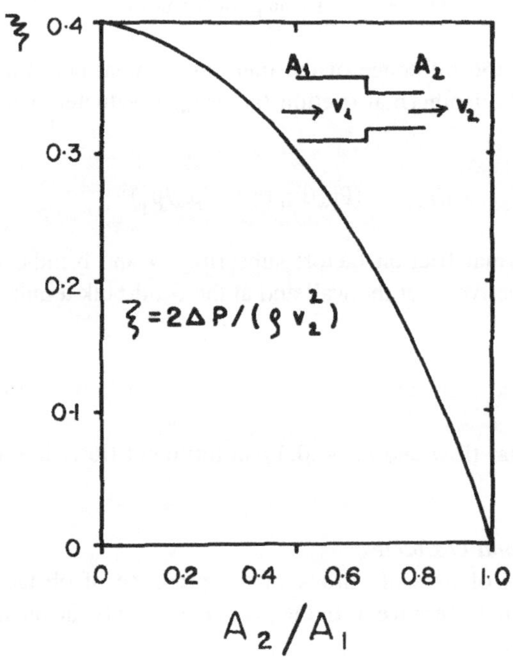

FIGURE 3. Concentrated loss coefficient (sudden contraction).

The dependence of ξ on Re is rather weak if the flow is turbulent; in such a situation, simplified diagrams are often used, where factor ξ is expressed only in function of (A_1/A_2) (Figures 2 and 3).

An analogous mechanism produces localized pressure drops in correspondence to tube bends. If the limited influence of Re is neglected and the flow is turbulent, the localized pressure drops may be expressed using the relationship:

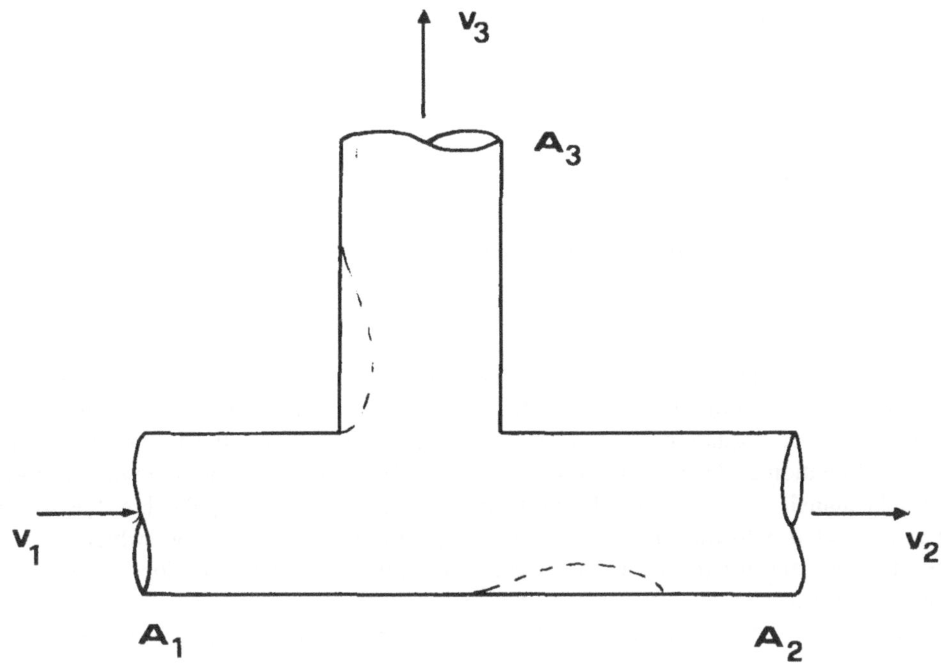

FIGURE 4. Single-phase flow through a tee.[7]

$$\Delta p_{obs} = \xi(R_c, \gamma)\ \rho\ v^2/2 \tag{11}$$

where ξ is given in function of the curvature radius R_c and of the curvature angle γ (followed by the axis of the tube in the bend)[4]:

$$\xi' = 0.051 + 0.19\ d_o/R_c \tag{12}$$

$$\xi = \begin{cases} \xi' & \text{if } \gamma = 90° \\ 0.9\ \xi' \sin\gamma & \text{if } \gamma < 70° \\ (0.7 + 0.35\ \gamma/90)\ \xi' & \text{if } \gamma > 100° \end{cases} \tag{13}$$

Equations 11 and 12 are valid for channel circular sections, with $R_c/d_o \geqslant 1$.

In the evaluation of pressure drops in a tee, the following relationship may be used[6]:

$$\Delta p_{1,i} = (\Delta p_{1,i})_{acc} + (\Delta p_{1,i})_{obs} \tag{14}$$

with $i = 2; 3$ (Figure 4), $(\Delta p_{1,i})_{acc}$ evaluated as in Equation 5, and:

$$(\Delta p_{1,i})_{obs} = \xi_{1,i}\ \rho_1\ v_1^2/2 \tag{15}$$

$$\xi_{1,2} = 0.1571 - 0.9197(G_3/G_1) + 1.0901(G_3/G_1)^2 \tag{16a}$$

$$\xi_{1,3} = 1.0369 - 0.9546(G_3/G_1) + 1.2123(G_3/G_1)^2 \tag{16b}$$

B. Flow Outside Tubes

Outside tubes, the flow is less confined and its local characteristics are less repeatable. This is particularly true in cross-flow configurations, with difficulties in experimental analyses of pressure drops and in the extrapolability of the results to different geometrical situations.

Correlations proposed for pressure drops outside tubes often make use, in their structure, of the results obtained in the analysis of pressure drops with flows inside tubes. Included in these are suitable correction terms which take into account, in a simple way, the particular geometry.

1. Parallel Flow

The evaluation of pressure drops in parallel flows is performed, in a first approximation, by considering the nontubular geometry through the use of the concept of the equivalent hydraulic diameter (Equation 3), and by experimentally evaluating the friction factor and employing Equation 2. Two limits exist for the use of such a model: the more the geometry of the subchannels is similar (ideal bank), the more Equation 3 is applicable; the second limit is connected with the "aleatoriety" of a typically-not-confined flow, which leads to a lack of agreement in expressions suggested in the literature for the friction factor.

With rod bundles (nuclear typology) of 6 and 37 rods, in circular geometry, that have values of $P_c/d_o = 1.57$ and $P_r/d_o = 1.46$ (P_c = circumferential pitch; P_r = radial pitch)[7]:

$$f = 0.05\ Re^{-0.2} \tag{17}$$

With reference to heat exchanger tube bundles[2] with triangular arrays $P/d_o = 1.46$, $2000 \leq Re \leq 10{,}000$, $3 \leq Pr \leq 6$, the friction factor may be expressed as:

$$f = a_1 Re^{a_2} + a_3 \tag{18}$$

where $a_1 = 22.5$, $a_2 = -1$, and $a_3 = 0.00205$.

The dispersion among experimental results and correlations available is due to the limited reproducibility of the experiences of flows inside tube banks (influence of nonideality factors in test sections), and is strongly dependent on the great number of parameters influencing the flow distribution as the interactions with the shell-wall, the ratio P/d_o, the kind and orientation of tube layout, the dimension of the tube bundle, the ratio between the internal diameter of the shell and the outside tube length, the presence of "disturbing" components such as grids, etc.

If the geometry is quite complex and a detailed analysis is required, use must be made of specific experimental data obtained in similar equipment. For finned tubes with transverse fins and a longitudinal flow of water ($1000 \leq Re \leq 150{,}000$ and $25 \leq T \leq 60°C$), with reference to the data of Table 1,[9] pressure drops may be evaluated through the relationship:

$$\Delta p = f_D (L/d_o)\ \rho\ v^2/2 \tag{19}$$

where f_D may be evaluated through Figure 5.

2. Cross Flow

The limits of the reproducibility of the flow and the difficulty of individuating a few parameters sufficiently representative of the local flow conditions, already pointed out for parallel flows outside tubes, increase considerably in cross flows. This is evidenced by the great variety of approaches followed by various researchers, which differentiate from one another even in the selection of the significative parameters.

Table 1
FINNED TUBE GEOMETRY PARAMETERS

Tube no.	d_f (mm)	$d_o \equiv d_r$ (mm)	l (mm)	B (mm)	s (mm)	A/Aext
1	38.0	16.49	1.00	0.85	1.49	2.48
2					2.99	1.74
3					5.98	1.37
4					11.95	1.19
5					23.90	1.09
6		16.27			∞	1.00

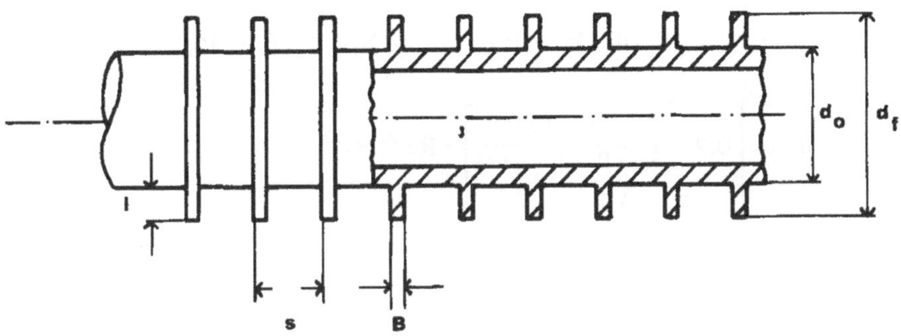

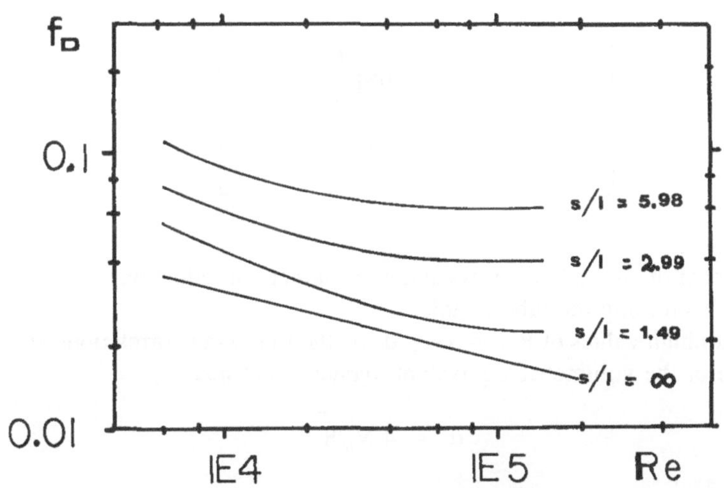

FIGURE 5. Best fit of friction factor $f_D = f_D$ (Re) (T = 25—60°C).

The geometry of the tube bank is considered to be a succession of localized section variations whose dissipative characteristics are influenced both by the turbulence level (Reynolds number) and by the distance between adjacent rows; that is (in usual configurations) the velocity profile between two adjacent rows not completely developed and in fast evolution.

As the subsequent tube rows are considered to be obstacles, use is often made, in the evaluation of pressure drops, of the following relationship structure:

$$\Delta p = 4 \, f \, n \, \rho \, v^2/2 \tag{20}$$

where n represents the number of tube rows in the direction of the flow, v is the velocity evaluated in a suitable flow area through the tubes, and the influence of the distance between adjacent rows and that of their array is included in the expression of f.

A rather used correlation is that proposed in Reference 10, which makes use of the following Reynolds number:

$$Re_o = G_{max}\, d_o/\mu \tag{21}$$

where G_{max} is the mass velocity evaluated in the minimum flowing section area (in correspondence to a plane passing through the axis of the bundle).

With $5000 \leqslant Re_o \leqslant 40,000$, the following expressions for n and for f are proposed:

- Staggered tube array:

$$n = N_f - 1 \quad \text{if} \quad P_t > P_{diag}; \quad n = N_f \quad \text{if} \quad P_t < P_{diag} \tag{22}$$

$$f = \left[0.25 + \frac{0.1175}{\left(\dfrac{P_t}{d_o} - 1\right)^{1.08}}\right] \cdot Re^{-0.15} \tag{23}$$

- In-line tube array:

$$n = N_f \tag{24}$$

$$f = \left[0.044 + \frac{0.08\left(\dfrac{P_l}{d_o}\right)}{\left(\dfrac{P_t}{d_o} - 1\right)^{\left[0.43 + 1.13\left(\frac{d_o}{P_l}\right)\right]}}\right] \cdot Re_o^{-0.16} \tag{25}$$

N_f being the number of crossed tube rows and n the number of reductions of area encountered by the flow while crossing the tube bundle.

Especially with high values of $Re_v = G_{max}\, d_v/\mu$, the following correlations are proposed[10] which make use of the volumetric equivalent diameter, defined as:

$$d_v = 4\, V_b/S \tag{26}$$

where V_b is the bundle free volume and S the heat transfer surface:

- For staggered array, P_{diag} being the pitch between two following rows measured along the conjunction line of two contiguous rows:

$$\Delta p = 4\, f \frac{G^2_{max}\, N\, P_l}{8\, \rho\, d_v} \cdot \left(\frac{d_v}{P_t}\right)^{0.4} \cdot \left(\frac{P_{diag}}{P_t}\right)^{0.6} \tag{27}$$

- For in-line array:

$$\Delta p = 4 \cdot f \frac{G^2_{max}\, N\, P_l}{8 \cdot \rho\, d_v} \cdot \left(\frac{d_v}{P_t}\right)^{0.4} \cdot \left(\frac{P_l}{P_t}\right)^{0.6} \tag{28}$$

where f is evaluated through the relationship:

$$f = 1.92 \cdot \left(\frac{G_{max}\, d_v}{\mu}\right)^{-0.145} \quad \text{if} \quad 5 \cdot 10^2 \leq \frac{G_{max}\, d_v}{\mu} \leq 2 \cdot 10^5 \tag{29}$$

With low values of Re_v, it is

$$f = \frac{180}{\left(\frac{G_{max}\, d_v}{\mu}\right)} \quad \text{if} \quad \frac{G_{max}\, d_v}{\mu} < 100 \tag{30}$$

For the evaluation of friction factor in finned tube banks with equilateral triangular tube layout, small fins and a number of fins per millimeter ranging between 0.4 and 1.06, it is proposed[11]:

$$\Delta p = 4\, f \cdot n\, \frac{G_{max}^2}{2\, \rho} \tag{31}$$

where f is defined as:

$$f = 3.805 \cdot Re^{-0.2336} \cdot \left(\frac{s}{d_f}\right)^{0.2512} \cdot \left(\frac{l}{s}\right)^{0.7593} \cdot \left(\frac{d_r}{d_f}\right)^{0.7292} \cdot \left(\frac{d_r}{P_t}\right)^{0.7085} \cdot \left(\frac{P_t}{P_l}\right)^{0.3791} \tag{32}$$

G_{max} is the mass velocity evaluated in the minimum free flow area, n is the number of area reductions, and the Reynolds number is defined as:

$$Re = d_r\, G_{max}/\mu \tag{33}$$

d_r is the tube diameter at the root of the fins; the other parameters are defined in Table 1. The relation is valid for fin height $l < 6.35$ mm, with Re ranging between 1000 and 25,000, $P_l < P_t$ and $n \leq 6$.

3. Geometry Variations and Obstacles

The limitations in the definition and use of correlations for flow outside tubes, previously mentioned are stressed when structure supports such as grids or baffles are present in the bundle. The grids, due to their complex geometry, cannot generally be simply assimilated to section variations or orifices, unless in first-approximation calculations. For the evaluation of pressure drops due to such components, the only method suitable for a detailed analysis is the use of empirical correlations obtained with the specific geometry. For spacing grids in parallel flows, the following relationship may be applied[12]:

$$\Delta p_a = \xi(Re)\, \rho\, v^2/2 \tag{34}$$

where v is the fluid velocity before the grid.

Figure 6 shows the data obtained experimentally[12] for the coefficient ξ, with Re ranging between 20,000 and 70,000 and an occupation factor of the grid of 25%. An extensive review of pressure drops analysis with spacing grids in square and triangular arrays may be found in Reference 13.

In cross flows, the influence on pressure drops of various geometric parameters representative of baffles, spacers, and supports may be found in Reference 16, while results on the effects of blockages and obstructions in heat exchangers with tube bundles may be found in Reference 14.

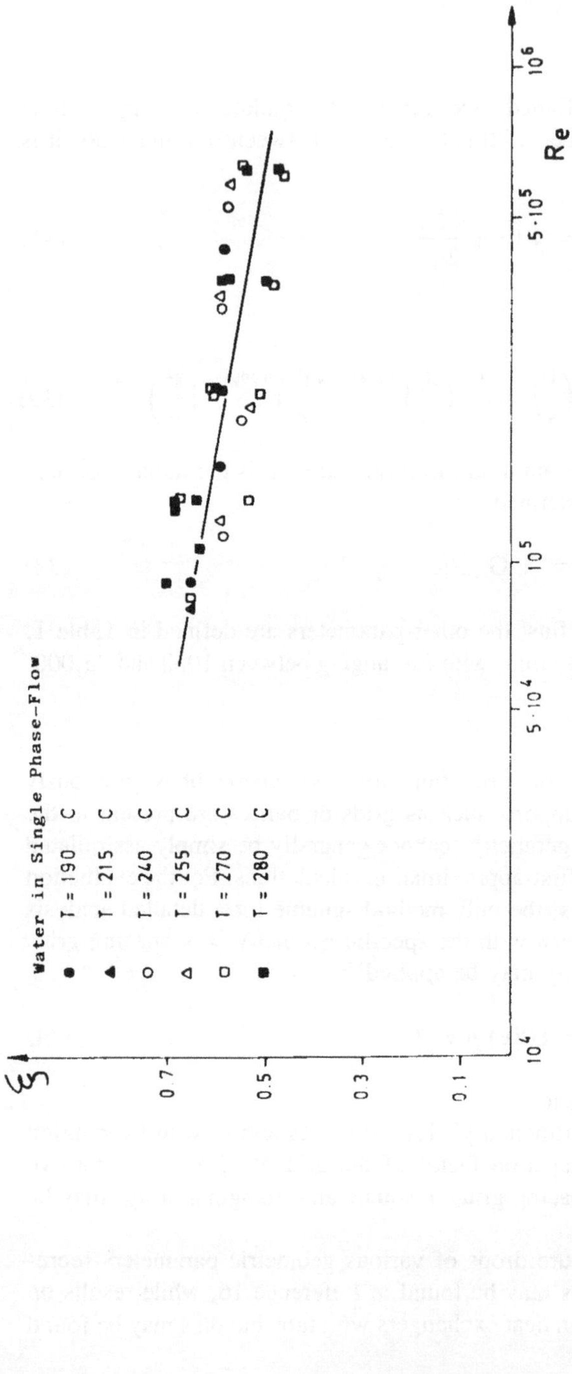

FIGURE 6. Experimental determinations of pressure drop coefficients for grids in BWR fuel bundles.

III. TWO-PHASE

A. Flow Inside Tubes

1. Adiabatic Flow Continuous Friction

As with single-phase flows, in two-phase flows, pressure drops in ducts (tubes, channels, other components) are generally expressed in the following form:

$$\Delta p_{tot} = \Delta p_f + \Delta p_{obs} + \Delta p_{acc} + \Delta p_{elev} \tag{35}$$

The third and the fourth terms of Equation 35 may be expressed as:

$$\Delta p_{acc} = \left\{\left[\frac{\rho_l X_2^2}{\alpha_2 \rho_g} + \frac{(1 - X_2)^2}{1 - \alpha_2}\right] - \left[\frac{X_1^2 \rho_l}{\alpha_1 \rho_g} + \frac{(1 - X_1)^2}{1 - \alpha_1}\right]\right\} \cdot \rho v^2 \tag{36}$$

$$\Delta p_{elev} = [\alpha \rho_g + (1 - \alpha) \rho_l] \cdot \Delta z \cdot g \tag{37}$$

In the computation of such terms relative to losses due to variations of the kinetic energy (Δp_{acc}) and to variations of the potential energy (Δp_{elev}), the evaluation of void fraction α is necessary; formulae and methods reported in Chapter 3 may be adopted.

The pressure losses due to the distributed friction and the nonreversible concentrated losses are generally expressed in the function of the corresponding single-phase pressure losses, making use of a multiplying coefficient ϕ^2. This is defined as the number that must be multiplied by the corresponding losses of a flow with the same mass flow rate, considered in saturated single-phase condition, in order to obtain the two-phase pressure drops (losses). Such an approach is based on the concept that the two-phase mixture may still be considered as a single-phase fluid with particular characteristics (for example, a "reduced" fluid density) and assumes implicitly that in each control volume the mixture characteristics are uniform.

A physical meaning may be attributed to such an approach only in the case of some flow regimes such as the bubbly flow or the dispersed, or, in general, when the geometry of the component to be analyzed is very large with respect to the dimensions of the local dishomogeneities of the two-phase flow. Other situations exist, nevertheless, where the concept of the two-phase multiplier has no real physical meaning, because the nature of the energy dissipation due to friction may be completely different in the two single phases. A typical example is that of an annular flow, where the friction losses are essentially due to the considerable superficial roughness encountered by the gaseous core current and to the annular liquid film flowing along the wall, with waves whose tips may be about five times the mean thickness of the film itself. In such situations, the method of the two-phase multiplier, not taking into account the physical behavior of the system, is approximated and may be useful only in gross evaluations. A sophisticated analysis requires, on the contrary, an approach considering the two phases as separated and interacting with each other. This last approach is much more complex because, in the various flow conditions, the interaction mechanisms between the two phases must be individuated and analyzed. Only recently, in the light of the needs imposed by the new set-up of the fluid systems conservation equations, has this approach been tried (see Chapter 2).

According to the two-phase multiplier method, the evaluation of the continuous friction and the concentrated nonreversible pressure drops may be obtained through the relationships*:

* Generally, in the description of pressure drops for a two-phase fluid, the single-phase pressure drops of reference are defined as if all the flow rate were liquid. In this situation, the suffix "lo" (liquid only) is used for the two-phase multiplier. Some authors refer to the single-phase pressure drops in the liquid phase as flowing alone; in this case, the use is made of suffix "l" or "lp" for the multiplier.

$$\Delta p_f = 4\ f\ \phi^2(X,\ p,\ G)\ \rho_l\ v_l^2(L/d)/2 \qquad (38)$$

$$\Delta p_{obs} = \xi\ \phi^2_{obs}(X,\ p,\ G)\ \rho_l\ v_l^2/2 \qquad (39)$$

The dependence of two-phase pressure drops on quality, pressure, and flow rate in the scale of the corresponding single-phase pressure drops is included in the multiplier.

On the basis of the preceding considerations, the functional dependence expected for the multiplier on the above parameters is rather complex. Several approaches have been tried based on experimental analyses in order to describe such dependence and extensive research is still in progress.

All these approaches may be conducted to two main models: homogeneneous and slip, differing in the hypothesis that the two phases may or may not have different velocities. The slip model is more complex, but is is better suited to the "intermediate" flow regimes (slug, etc.; see Chapter 3). The homogeneous model is applicable satisfactorily to flows with extreme void fractions (very high or very low; bubbly flow, dispersed flow), high pressure, and high mass velocities.

a. Homogeneous Model

The basic hypotheses of the homogeneous model are

1. Same linear velocity for the gaseous and the liquid phases
2. Thermodynamic equilibrium between the two phases

It is assumed that the phases are homogeneously mixed and the dissipative mechanism of interaction with the wall is similar to that of a single-phase fluid, so that an analogous friction factor may be considered.

The physical properties of the two-phase flow are obtained from the properties of the single phases:

$$\Delta p_f = 4\ f_{tp}\ \frac{G^2}{2\bar{\rho}} \cdot \frac{L}{d} \qquad (40)$$

where:

$$1/\bar{\rho} = X/\rho_g + (1 - X)/\rho_l,\ \text{i.e.,}\ \rho = \alpha\ \rho_g + (1 - \alpha)\ \rho_l \qquad (41)$$

$$f_{tp} = f(Re_{tp}) \qquad (42)$$

f_{tp} being evaluated using single-phase friction factor correlations and α evaluated through Equation 11, with $S = 1$.

Several alternative forms for the definition of the two-phase Reynolds number Re_{tp} have been proposed, varying in the definition of the two-phase viscosity (see Chapter 1). Among them, the best results are obtained with the assumption of:

$$Re_{tp} = G\ d/\mu_l$$

b. Slip Model

The slip model has been proposed in its original form by Lockhart and Martinelli[15] who introduced an adimensional parameter X_{tt} (Chapter 1, Section IV.A 16) and a two-phase multiplier refering to gas/liquid single-phase (flowing alone) pressure drops.

The Lockhart-Martinelli method, obtained with reference to low-pressure air/water systems, has been extensively used in the past; presently, other more accurate methods are preferred. One of the best of these, making use of Martinelli parameter X_{tt} and referring only to the liquid phase (supposed flowing alone), is here suggested especially with intermediate values of X_{tt} defining the two-phase friction multiplier as[16]:

$$\phi_l^2 = 1 + \frac{C}{X_{tt}} + \frac{1}{X_{tt}} \tag{43}$$

where:

$$C = \left[1 + (C_2 - 1) \cdot \left(1 - \frac{\rho_g}{\rho_l}\right)^{0.5}\right] \cdot \left[\left(\frac{\rho_l}{\rho_g}\right)^{0.5} + \left(\frac{\rho_g}{\rho_l}\right)\right]^{0.5} \tag{44}$$

The parameter C_2 is a function of the pipe characteritics:

- Rough tubes:

$$C_2 = 1 \quad \text{for} \quad G > 1500 \, \frac{\text{kg}}{\text{m}^2\ \text{sec}} \quad \text{or} \quad 9 \leqslant \left(\frac{\rho_l}{\rho_g}\right)^{0.5} \leqslant 30 \tag{45}$$

$$C_2 = \frac{1500}{G} \quad \text{for} \quad G < 1500 \, \frac{\text{kg}}{\text{m}^2\ \text{sec}} \quad \text{and} \quad \left(\frac{\rho_l}{\rho_g}\right)^{0.5} < 9 \tag{46}$$

- Smooth tubes:

$$C_2 = 1 \quad \text{for} \quad G > 2000 \, \frac{\text{kg}}{\text{m}^2\ \text{sec}} \quad \text{or} \quad 9 \leqslant \left(\frac{\rho_l}{\rho_g}\right)^{0.5} \leqslant 30 \tag{47}$$

$$C_2 = \frac{2000}{G} \quad \text{for} \quad G < 2000 \, \frac{\text{kg}}{\text{m}^2\ \text{sec}} \quad \text{and} \quad \left(\frac{\rho_l}{\rho_g}\right)^{0.5} < 9 \tag{48}$$

Equation 43 is applicable not only for the evaluation of the distributed, but also for the evaluation of the concentrated friction pressure losses, with a suitable choice of parameter C_2.

Varying the value of parameter C_2, the Lockhart-Martinelli and the Baroczy curves may be obtained. The structure of Equation 43, furthermore, is such that it adequately interprets the various transitions among the possible flow regimes. For this reason, it is also used in the evaluation of the influence of quality on two-phase heat transfer coefficients.

Even if the described model is rather complete, when relative drag interactions between phases or buoyancy forces are particularly relevant (vertical downflow, annular upflow, separate phase flow), the following model is suggested. The model makes use of a set of correlations for the two-phase friction multiplier,[17,18] referring to 25,000 experimental data (with water-steam systems, Refrigerant 12 systems, water-air systems, and oil-air systems) to be used with horizontal and vertical geometrics, with circular, rectangular, and annular sections (adiabatic flows):

- Horizontal and vertical upflow:

$$\phi_{lo}^2 = A + \left[3.24 \cdot X^{0.78} \cdot (1 \cdot X)^{0.224} \cdot \left(\frac{\rho_l}{\rho_g}\right)^{0.91} \cdot \right.$$

$$\left. \cdot \left(\frac{\mu_g}{\mu_l}\right)^{0.19} \cdot \left(1 - \frac{\mu_g}{\mu_l}\right)^{0.7}\right] \Big/ [Fr_{tp}^{0.0454}\ We_{tp}^{0.035}] \tag{49}$$

- Downflow:

$$\phi_{lo}^2 = A + \left[48.6\ X^{0.8} \cdot (1 - X)^{0.29} \cdot \left(\frac{\rho_l}{\rho_g}\right)^{0.9} \cdot \left(\frac{\mu_g}{\mu_l}\right)^{0.73} \cdot \right.$$

$$\left. \cdot \left(1 - \frac{\mu_g}{\mu_l}\right)^{7.4} \cdot Fr_{tp}^{0.03}\right] \Big/ [We_{tp}^{0.12}] \tag{50}$$

with:

$$A = (1 - X)^2 + X^2\left(\frac{\rho_l\ f_g}{\rho_g\ f_l}\right) \tag{51}$$

$$\rho_{tp} = \left(\frac{X}{\rho_g} + \frac{1 - X}{\rho_l}\right)^{-1} \tag{52}$$

$$\mu_{tp} = \left(\frac{X}{\mu_g} + \frac{1 - X}{\mu_l}\right)^{-1}; \quad Fr_{tp} = \frac{G^2}{(g\ d_{eq}\ \rho_{tp}^2)}; \quad We_{tp} = \frac{G^2\ d_{eq}}{(\sigma\ \rho_{tp})} \tag{53}$$

The evaluation of single-phase pressure drops is performed through Equation 2, where f is evaluated as follows:

- Circular sections:

$$f = \frac{16}{Re} \quad \text{for} \quad Re \leq 1055 \tag{54}$$

$$f = \frac{1}{4}\left[0.86859\ \ln\left(\frac{Re}{1.964\ \ln Re - 3.8215}\right)\right]^{-2} \text{for } Re > 1055 \tag{55}$$

- Rectangular sections:

Equation 54 and 55 are used, where Re must be substituted by another adimensional parameter defined as:

$$Re^* = \psi \cdot Re \tag{56}$$

being:

$$\psi = 2/3 + (11/24)(s/w)\ (2 - s/w) \quad s \leq w \tag{57}$$

where s and w are the two side dimensions.

- Annular sections:

$$f = \frac{16}{Re} \quad \text{for} \quad Re \leq 1055 \tag{58}$$

$$f = \frac{1}{4}\,[2\,\log_{10}(Re \cdot \sqrt{4f} - E]^{-2} \quad \text{for} \quad Re > 1055 \tag{59}$$

where:

d_i/d_o	0	0.05	0.3	0.6	1.0
E	0.8	0.932	0.961	0.968	0.97

A method which takes into account the occurring flow regime (see Chapter 3) may be found in Reference 19 and is suggested in analyses of two-phase systems both in adiabatic and in diabatic conditions (it requires the knowledge of the flow-regime).

2. *Diabatic Flow Continuous Friction*

The two-phase pressure drops models reported in the previous section are obtained and proposed (with the exception of the last model) with reference to adiabatic conditions; nevertheless, they are often used for diabatic flows as well. Some models have also been developed with specific reference to diabatic conditions, mainly aimed at the analysis of the effects of specific phenomena such as incipient boiling, mass boiling, and condensation on the two-phase pressure drops. They are characterized by a generally limited range of application.

The influence of a contemporary heat transfer on two-phase flow pressure drops is due (especially in boiling) to the modification, with respect to the corresponding adiabatic situation, in the velocity profiles of each phase and in the spatial distribution of voids.

a. Local Boiling Effect

The effect of local boiling (subcooled boiling) is particularly important in pressurized water nuclear reactors.[21] Such an effect varies with pressure.[20,21]

b. Bulk Boiling Effect

For the evaluation of two-phase pressure drops in bulk boiling, see Reference 21.

c. Condensation Effect

The pressure drops of a condensing flow may be evaluated through adiabatic correlations which make use of local flow parameters; this method leads to acceptable results only in a first approximate evaluation.

A specific method[22] for the evaluation of the friction component of two-phase pressure drops with a condensing fluid is reported here, making reference to adiabatic conditions, but also taking into account the effect of mass transfer on the vapor-liquid interface:

$$\Delta p_f = \Delta p_{ad} \cdot \phi \tag{60}$$

where Δp_{ad} are the adiabatic-condition pressure drops and ϕ is obtained as:

$$\phi = \theta/[1 - \exp(-\theta)] \tag{61}$$

where:

$$\theta = G_{cond}(v_v - v_i)/\tau_i \tag{62}$$

θ takes into account the effect of mass transfer at the vapor-liquid interface, G_{cond} is the condensing mass velocity, v_v and v_i are the axial velocities, respectively, of the vapor phase and of the interface, and τ_i is the shear stress which results at the interface. The knowledge of interface mechanical interactions is clearly necessary.

3. Geometry Variations and Obstacles

In analogy to friction pressure drops, the pressure drops due to variations in the channel geometry or to obstacles are generally evaluated making use of the corresponding correlations for single-phase flow and of two-phase multipliers. The most common hypotheses at the basis of the proposed models, are those of a one-dimensional flow and of nonvariability of the phases while passing through the geometry variation or crossing the obstacle. For a generic abrupt variation of the geometry, the following relationship may be used:

- Homogeneous model:

$$\phi^2_{lo,obs} = 1 + X[(\rho_l/\rho_g) - 1] \tag{63}$$

- Slip model[16]:

$$\phi^2_{l,obs} = 1 + C/X_{tt} + 1/X^2_{tt} \tag{64}$$

C being evaluated through Equation 44, where parameter C_2 assumes the following values: increase in the duct diameter: 0.5; reduction in the duct diameter: 1.0; tee: 1.75, 90° bends: $1 + 35$ d/L, ($L = \pi R_c/2$ = bend length, R_c = curvature radius); gates: 1.5; and ball valves: 2.3. For a deeper analysis, see References 23 to 25.

B. Flow Outside Tubes

The nonreversible pressure drops with two-phase flows outside tube bundles are generally expressed in analogy to flows inside tubes, making reference to the corresponding single-phase pressure drops with the introduction of a multiplier. The general expression is, therefore, of the kind:

$$\Delta p_{tp} = \phi^2 \Delta p_{sp} \tag{65}$$

where multiplier ϕ^2 depends mainly on the flow regime (cross, parallel; slug, annular, etc.), the flow orientation, and the macroscopic fluidodynamic parameters of the two-phase flow such as quality, void fraction, pressure, etc.

In general, the experimental analyses have concerned flows with constant quality and in ideal geometries (absence of bypasses) which are sometimes quite far from the real situations of heat exchange apparatuses. The "real" situation may be taken into account using non-ideality factors (multiplying factors, experimentally evaluated).

If a consistent variation occurs in the characteristics of the two-phase flow inside the bundle (due to pressure losses and/or to heat transfer), the procedure of step-by-step calculations is suggested.

1. Parallel Flow

A correlation[26] is suggested, also valid for two-phase pressure drops inside tubes:

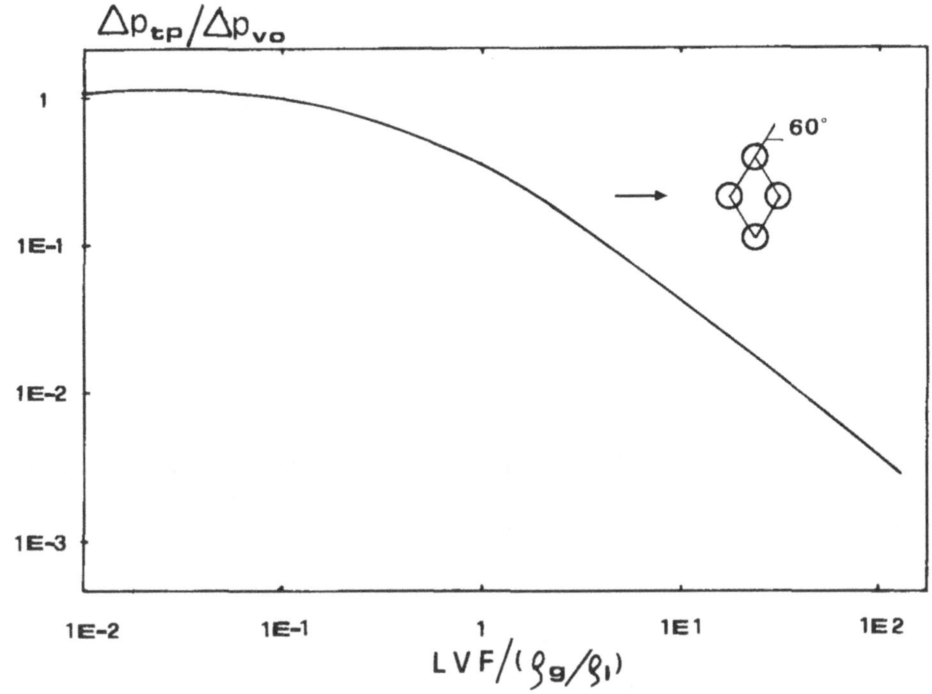

FIGURE 7. Two-phase cross flow pressure drops through a horizontal tube bundle.

$$\Delta p = \frac{a_1 \, G^{a_2} \, v^{0.86} \, \sigma^{0.4} \, L}{d_{eq}^{1.2}} \tag{66}$$

where d_{eq} is the equivalent hydraulic diameter, $a_1 = 0.0354$, $a_2 = 1.6$, and:

$$v = X/\rho_g + (1 - X)/\rho_l \tag{67}$$

The experimental ranges are $500 \leq G \leq 5000$ kg/(m^2 sec); $0.005 < d_{eq} < 0.025$ m; $0.1 < L < 4$ m; $0.02 < \sigma < 0.08$ kg/sec^2; $15 < \rho_l/\rho_g < 100$; $0.01 < X < 1$.

Other correlations developed for pressure drops in parallel flow are reported in the following section as a part of studies mainly oriented to the analysis of cross flows.

2. Cross Flow

Cross flow pressure drops with horizontal flows in tube bundles in adiabatic conditions may be evaluated through Figure 7,[27] obtained with experiences in four experimental sections, with $Re_v = G\,d/\mu_v \geq 5000$ and with mass velocity G evaluated in the minimum cross area. The diagram in Figure 7 allows the evaluation of Δp_{tp} with reference to single-phase pressure drops, considering the whole flow as vapor (Δp_{vo}). In abscissa, the ratio:

$$LVF/(\rho_g/\rho_l) = (1 - X)/X \tag{68}$$

is reported, in which LVF (liquid volume fraction) is evaluated using the homogeneous model.

Analogous diagrams relative to different tube arrays may be found in Reference 27. In the same reference, the authors suggested the use of the diagrams in the evaluation of pressure drops through a tube bundle with a cross flow of condensing vapor. It is advisable, however, to subdivide the bundle in zones where quality may be considered as constant.

Table 2
CONSTANTS a_1 AND a_2 IN EQUATION 69

Cross-flow zone	a_1	a_2
Flow regime		
Vertical, up- and downflow, spray, and bubbly flow	0.37	1.0
Horizontal side-to-side spray and bubbly flow	0.46	0.75
Horizontal side-to-side stratified and stratified-spray flow	0.46	0.25
Parallel zone	a_1	a_2
Flow regime		
Vertical, up- and downflow	0	$(\rho_l/\rho_{tp})^{0.25}$
Horizontal side-to-side flow	0	$2/(\omega + 1)$

An analysis specifically aimed at pressure drops in ideal heat exchangers (Chapter 3, Figure 9) is contained in Reference 28, where the cross flow and the window flow zones are distinguished. The relationship proposed is

$$\frac{\Delta p_{tp}}{\Delta p_{lo}} = 1 + (\omega^2 - 1) \cdot \left[a_2 \, X^{\left(\frac{2 - a_1}{2}\right)} \cdot (1 - X)^{\left(\frac{2 - a_1}{2}\right)} + X^{(2 - a_1)} \right] \tag{69}$$

with:

$$\omega^2 = \Delta p_{vo}/\Delta p_{lo} \tag{70}$$

Values for parameters a_1 and a_2 are shown in Table 2, where ρ_{tp} and α are evaluated using the homogeneous model (through, respectively, Equations 24 and 11 and from Chapter 1, where $S = 1$).

C. Advanced Methods

The development of the treatment of two-phase mixtures in terms of systems with two distinct phases flowing contemporaneously inside the same control volume has made it necessary (as pointed out in Chapter 2) to overcome the classical approach of the only fluid-wall mechanical interaction, based on the use of Equation 1 and its extensions. It becomes necessary, in fact, to describe the modes of the mechanical interaction of each phase with the solid wall of the control volume and with the reciprocal phase (see separate-phase models, Chapter 2).

This description of the two-phase mixture is quite complex, owing to the difficulty in the experimental evaluation of such interactions and because it imposes the development of different models for the various flow regimes. Each flow condition, in fact, is the consequence of the equilibrium of the actions (both thermal and mechanical) exchanged by the phases and with the wall, whose importance varies with the flow regime.

In order to use the separate-phase analysis method in the pressure drops evaluation, a loss coefficient per unit area of contact between the phases and a contact area of each phase with the wall and of each phase with the other per unit volume of the mixture, must be introduced. Such parameters must be defined with reference to each flow regime. In the loss coefficient definition, both the effect of the viscous interaction and of inertia at the interface are generally included. The loss coefficient, F_{lg}, and the contact area between the phases per unit volume are well known for bubbly flow (case I: $\alpha < 0.3$) and annular flow (case II: $\alpha \geq 0.8$).

1. Case I (α <0.3)

In the bubbly flow regime:

$$F_{lg} = \frac{3}{4} C_D \frac{(1 - \alpha)}{d_b} \alpha(1 - \alpha)^3 \rho_l(v_g - v_l) |v_g - v_l| +$$

$$+ C\rho_l v_g \alpha \frac{d(v_g - v_l)}{dz} \tag{71}$$

$$A = N \pi d_b \tag{72}$$

where d_b is the bubble diameter and N the number of bubbles per unit volume. The second term on the right-hand side of Equation 71 is very important in highly accelerated flows (i.e., critical flows), because it takes into account the variations of the kinetic energy induced by relative motion of the phases.

Assuming the bubbles as insulated rigid spheres ($0 \leq \alpha \leq 0.2$), the coefficients C and C_d may be evaluated as follows[29]:

$$C = 0.5 \tag{73}$$

$$C_D = (C_D)_R = \begin{cases} 0.44 & \text{for } Re_b > 1000 \\ \dfrac{24}{Re_b} \cdot (1 + 0.15\, Re_b^{0.687}) & \text{for } Re_b < 1000 \end{cases} \tag{74}$$

where Re_b is the bubble Reynolds number:

$$Re_b = \rho_l d_b(v_g - v_l)(1 - \alpha)/\mu_l \tag{75}$$

If it is considered that the bubbles may interfere with one another ($0.2 \leq \alpha \leq 0.3$), the drag coefficient has to be modified as follows:

$$C_D = (C_D)_R (1 - \alpha)^{-4.7} \tag{76}$$

2. Case II (α >0.8)

$$F_{lg} = 3 \frac{f_{lg}}{d} \sqrt{\alpha}\, \rho_g(v_g - v_l) |v_g - v_l| \tag{77}$$

$$A = 4\, \alpha^{0.5}/d \tag{78}$$

where d is the tube internal diameter and f_{lg}, interfacial friction factor, is

$$f_{lg} = 0.005\, [1 + 75(1 - \alpha)] \tag{79}$$

3. Intermediate Region (0.3 < α < 0.8)

As a first approximation, in the intermediate region (0.3 < α <0.8), it may be assumed[29] for the interfacial friction factor a linear interpolation through the void fraction α:

$$f_{lg} = \frac{[f_{lg}]_b - [f_{lg}]_a}{(\alpha_b - \alpha_a)}(\alpha - \alpha_a) \tag{80}$$

where α_b is the void fraction where coalescence of bubbles starts, $[f_{lg}]_b$ is the interfacial friction factor at this particular void fraction, and $[f_{lb}]_a$ is the interfacial friction factor at the onset of annular flow at the void fraction α_a (α_a = 0.8).

REFERENCES

1. **Reactor Handbook,** Vol. 4, John Wiley & Sons, New York, 1964.
2. **Sieder, E. N. and Tate, G. E.,** Heat transfer and pressure drop of liquids in tubes, *Ind. Eng. Chem.*, 28(12), 1936.
3. **Idel'cik,** *Memento des pertes de charge,* Eyrolle, Paris, 1968.
4. **Nekrassov, B.,** *Cours d'Hydraulique,* MIR, 19.
5. **Perry, R. H. and Chilton, C. H.,** *Chemical Engineers Handbook,* 5th ed., McGraw Hill, New York, 1973.
6. **Reimann, J. and Seeger, W.,** Pressure drop of air water and steam water flow in a dividing T junction, paper presented at Eur. Two-Phase Flow Group Meet., June 19, Rome, 1984.
7. **Nylond, O.,** Frigg-3, AB Atomenergyi and ASEA-ATOM.
8. **Wantland, J. L.,** Compact Tubular Heat Exchangers, USAEC Rep. TIC 7529, U.S. Atomic Energy Commission, Washington, D.C., 1957.
9. **Obermeier, E. and Schaber, A.,** Experimental investigation of heat transfer, paper presented at 6th Int. Heat Transfer Conf., Toronto, 1978.
10. **Cumo, M., Farello, G. E., Iovino, F., and Naviglio, A.,** Single and Two-Phase Cross-Flow Through Tube Bundles, CNEN RT/ING (79), 30, 1979.
11. **Rabas, T. J., Eckels, P. W., and Sabatino, R. A.,** The effect of fin density on the heat transfer and pressure drop performance of low-finned tube banks, paper presented at ASME-AIChE Natl. Heat Transfer Conf., Orlando, Fla., July 27, 1980.
12. **Grillo, P. and Mazzone, G.,** Single and two-phase pressure drops on 6 × 6 rod bundle at 70 atm, *Nucl. Technol.*, 15, July 1972.
13. **Rehme, K.,** Pressure drop correlations for fuel element spacers, *Nucl. Technol.*, 17, January 1972.
14. **Sparrow, E. M. and Ruiz, R.,** Effect of blockage-induced flow maldistribution on the heat transfer and pressure drop in a tube bank, *J. Heat Transfer,* November 1982.
15. **Lockhart, R. W. and Martinelli, R. C.,** Proposed correlation of data for isothermal two-phase two-component flow in pipes, *Chem. Eng. Prog.*, 45, 39, 1949.
16. **Chisholm, D. and Sutherland, L. A.,** Prediction of pressure gradients in pipeline systems during two-phase flow, *Proc. Inst. Mech. Eng.*, 184(3C), 1969—1970.
17. **Friedel, L.,** Improved friction pressure drop correlations for horizontal and vertical two phase pipe flow, paper presented at Eur. Two-Phase Flow Group Meet., Ispra, June 5, 1979.
18. **Friedel, L.,** Two-phase friction pressure drop correlation for vertical downflow, paper presented at Eur. Two-Phase Flow Group Meet., Rome, June 19, 1984.
19. **Beattie, D. R. M.,** A note on the calculation of two-phase pressure losses, *Nucl. Eng. Design,* 25, 1973.
20. **Reynolds, J. B.,** Local Boiling Pressure Drop, USAEC Rep. ANL 5178, U.S. Atomic Energy Commission, Washington, D.C., 1954.
21. **Tong, L. S.,** *Boiling Heat Transfer and Two-Phase Flow,* John Wiley & Sons, New York, 1965.
22. **Owen, R. G., Sardes, R. G., Butterworth, A.,** Two-phase pressure drop for condensation inside a horizontal tube, paper presented at Int. Semin. Advancement in Heat Exchangers, Dubrovnik, Yugoslavia, September, 1981.
23. **Lottes, P. A.,** Expansion losses in two-phase flow, *Nucl. Sci. Eng.*, 9, 26, 1961.
24. **Janssen, E.,** Two-phase pressure loss across abrupt contractions and expansions steam water at 600-1400 psia, Int. Heat Transfer Conf., Vol. 5, American Society of Mechanical Engineers, 1966.

25. **Hewitt, J. W.,** Two-phase flow measurements with orifices, *Trans. ASME Ser. D: Basic Eng.*, 84, 1962.
26. **Grassi, Lombardi, and Pedrocchi,** Cadute di pressione in flusso bifase, CISE R-288, 24th Congr. Natl. ATI, Bari, October, 1969.
27. **Diehl, J. E. and Unruh, C. H.,** Two Phase Pressure Drop for Horizontal Cross-Flow Through Tube Banks, ASME Pap. 58-HT-20,9, American Society of Mechanical Engineers, New York, 1957.
28. **Grant, I. D. R. and Chisholm, D.,** Two phase flow on the shellside of baffled shell and tube heat exchanger, *Trans. ASME*, February 1979.
29. **Richter, H. J.,** Separated two-phase flow model: application to critical two-phase flow, *J. Multiphase Flow*, 9(5), 1983.

Chapter 5

SINGLE-PHASE NATURAL CONVECTION

I. INTRODUCTION

The term "natural convection" denotes the phenomena characterized by a fluid motion which is due only to heat transmission. While in forced-convection problems it is often possible to separate the thermal from the fluidodynamic analysis (the last one being fundamentally dependent on the characteristics of the fluid motor), in natural convection problems the two analyses must be performed coincidentally.

In this chapter, only single-phase natural convection phenomena are taken into consideration. For better comprehension of them, a horizontal plane slab may be considered, invested by a parallel fluid flow, as in Figure 1.[1] Initially, the hypothesis of an external forcing action of the fluid motion is done (forced convection). The presence of the fixed slab slows down the fluid layer adjacent to it; in turn, such a layer (through viscous actions) slows down fluid layers farther and farther from the solid slab.

The kinematic boundary layer is defined as a zone which feels the effects of the presence of the slab and is generally identified as the fluid layer between the slab and a distance from it such as the fluid velocity (which has a null value on the wall, in the absence of slip) has a value equal to 99% of the undisturbed fluid velocity (U_∞). Such a layer depends strongly on the fluid viscosity; the mechanical interaction fluid/slab is concentrated in it, the other part of the fluid remaining undisturbed.

Furthermore, if the slab is at a different temperature with respect to the fluid, a thermal boundary layer is also defined which is the fluid part thermally affected by the slab at different temperature; it has such a thickness that the temperature in it passes from the wall value (T_w) to the undisturbed-fluid value (T_∞). Generally, the thickness of the two layers is different, except under certain conditions. In the situation of natural convection, the phenomenology is substantially different from the case of Figure 1 (forced convection).

To point out the difference a vertical slab of limited height H is considered maintained at a constant temperature T_w, and submerged in an infinite fluid at temperature T_∞ (Figure 2). The slab heats the adjacent fluid layers and provokes their expansion. The buoyancy (Archimede) forces act on the fluid volumes at different temperatures (and, therefore, with different densities), causing an upward motion of the heated fluid, which leaves space to colder fluid coming from the bottom. Natural circulation of the fluid is therefore initiated. Fluid circulation may be regarded as a cycle of a thermal machine operating between two temperatures, T_∞ and T_w; along the path of a small fluid mass, the following processes occur: heating/expansion/cooling/compression.

Clearly, mechanical energy is available as an effect of the fluid motion; fluid is accelerated up to a velocity such that, under stationary conditions, the buoyancy forces are equal to the friction forces among the fluid currents. The difference between forced and free (natural) convection is, therefore, clear: first, the motor is external to the fluid; second, it is internal to the fluid and most of all depends on the temperature distribution. In natural convection, the motion is initiated by the density gradients in the fluid due to the nonuniform temperature field; temperature is the driving force. It is also important to emphasize that such thermal-convective flows could not exist if the effect of a force field, such as that of gravity, were not present.

In a (vertical) gravity field, with vertical heating slabs, flow is also possible with low ΔT. Should a slab be in a horizontal position and with a fluid in contact with its upper side, the flow of the fluid would be possible only with ΔT higher than a critical value (Ra >1100).

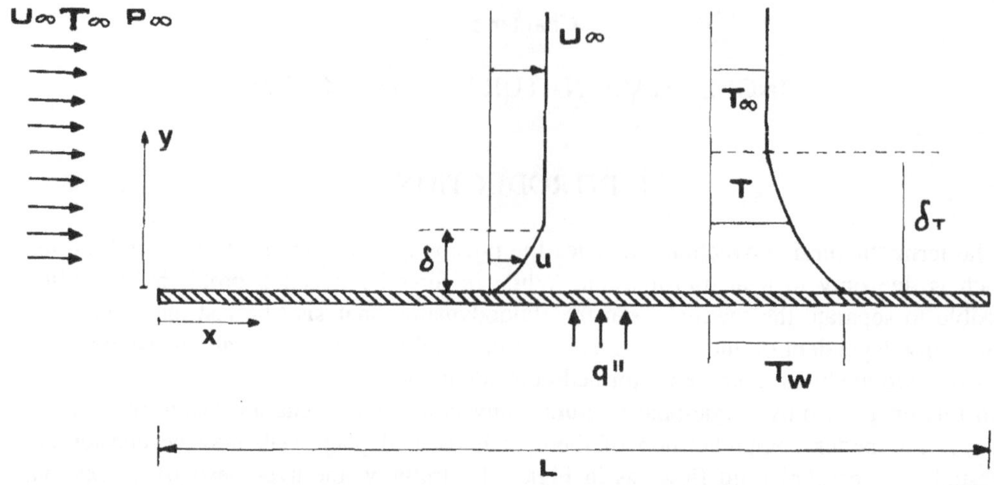

FIGURE 1. Velocity and thermal boundary layers.

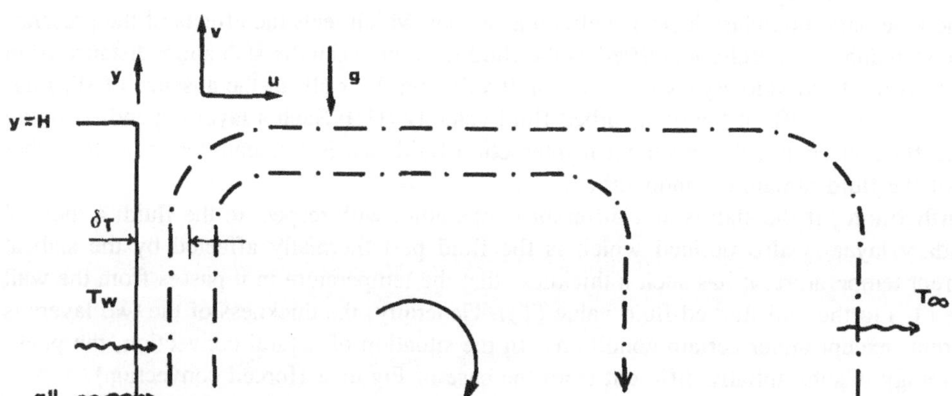

FIGURE 2. Natural convection due to a heating vertical slab.

In such a situation, in fact, the warm fluid, moving upward, serves as an obstacle to the downward flow of the cold fluid. The buoyancy force, generated by the temperature gradients, has to be strong enough to break the warm fluid layer in contact with the slab. Typical examples of natural convection are in rooms, in the presence of radiating panels; warm fume flows in the atmosphere; and cooling fins in electronic components.

In all these situations, the main problem to be solved is the determination of one or more of the following quantities:

1. Heat transferable to the fluid when the wall temperature is imposed
2. Maximum wall temperature when heat flux on the wall is imposed
3. Forces exerted by the fluid on the wall

Other natural convection phenomena occurring in closed piping systems (natural circulation loops, thermosyphon plants, nuclear primary cooling circuits in pump-breakdown condition) require an analysis using methods of forced convection.

II. FUNDAMENTALS

A. Velocity and Temperature Profiles

Within the limits of this text, reference is made only to plane geometries (heating/cooling slabs); the extension to other geometries is often easy. With reference to a horizontal slab (as in Figure 1), the following quantities must be generally evaluated in terms of a natural-convection problem: the force exerted by the fluid on the slab and the heat transferred from the slab to the fluid. The exerted force is viscous in nature and may be represented as:

$$F = \int_0^L \tau \, z \, d \, x \tag{1}$$

where Z is the orthogonal dimension with respect to the figure. For the evaluation of transferred heat:

$$q = \int_0^L q'' \, z \, d \, x \tag{2}$$

The evaluation of τ and q'' is therefore necessary.

The Newton law may be assumed:

$$\tau = \mu(\partial u/\partial y)_{y=0} \tag{3}$$

which permits (if the velocity profile is known) the evaluation of τ. For the evaluation of q'', the hypothesis may be made (supported by experimental evidence) of a motionless fluid sublayer in contact with the wall (no slip). Through the motionless sublayer, heat may be transferred only for conduction:

$$q'' = -k(\partial T/\partial y)_{y=0} \tag{4}$$

If Equations 3 and 4 are analyzed, it is easy to understand why all the researchers' efforts are focused on the evaluation of velocity and temperature profiles: they allow, in fact, the determination of local velocity and temperature gradients from which τ and q'' are obtained.

B. Solving Equations

For the evaluation of the physical parameters (velocity and temperature profiles) which allow the determination of the fluid stress on the wall and that of the exchanged heat, the continuity equation, momentum equations and the energy equation must be used (Chapter 2).

If a linear dependence of the viscous stress on the velocity gradient may be hypothesized (Equation 3), if μ = const (Newtonian fluid) and if ρ = const along the direction of flow, the Navier-Stokes equations (as a particular form of momentum equations) may be used (Chapter 2, Equation 9). The contemporary solution of continuity (one equation), momentum (three equations), and energy (one equation) equations allows the evaluation of the five unknown parameters describing the hydraulic and thermal behavior of the fluid: u, v, w, p, T. Such equations are generally quite difficult to solve and many approximations are often necessary; the problem of determining the shear stress on the wall and the heat transfer to and from the fluid in natural convection is, finally, transferred to that of selecting the most suitable method of solution of conservation equations.

An approximated method (integral method) used for the resolution of such equations in natural convection problems is described in Section D; another (similarity method) is described in Reference 1. Resolutive methods are generally used after the application of a scale analysis (see next section).

C. Scale Analysis

This method is very helpful in the resolution of several heat transfer problems, especially pertaining to natural convection, giving the order of magnitude of the physical parameters of interest. A selection of the most relevant parameters for the specific problem follows, leading to a simplification of the problem itself. The method of scale analysis has an importance which is often not sufficiently stressed. Scale analysis must not be mistaken for the dimensional analysis: the latter verifies if the two members (or their addenda) of an equation are physically homogeneous, while the scale analysis has the aim of determining, among the terms of an equation, which of them are the most relevant, in order to perform simplificative hypotheses on the phenomenon.

Using this method, it is also possible to obtain relationships predicting — less a multiplicative factor of order 1 — the final exact solutions. The "real" solutions must then be researched using analytical or numerical methods for the resolution of the differential equations; such methods, however, use the scale analysis because, thanks to the first simplificative hypotheses introduced, they may be considerably simplified.

As an example of the application of the scale analysis method, the thickness of the boundary layer of a fluid is found flowing parallel to a slab with a velocity U_∞ and in the following hypotheses (see Figure 1): constant viscosity, stationary flow, incompressible fluid, negligible viscous stresses in y direction, and negligible pressure gradients in x direction.

With such hypotheses, the momentum conservation equation (with reference to axis x; see Chapter 2, Equation 9a) and the continuity equation become:

$$u\,\partial u/\partial x + v\,\partial u/\partial y = \nu\,\partial^2 u/\partial y^2 \qquad (5)$$

$$\partial u/\partial x + \partial v/\partial y = 0 \qquad (6)$$

The terms $u\partial u/\partial x$ and $v\partial u/\partial y$ represent the inertia of the fluid, while the term $v\partial^2 u/\partial y^2$ represents the friction force. Furthermore, since only the boundary layer region is considered, it is possible to substitute u (and ∂u) with U_∞ and y (and ∂y) with δ (boundary layer thickness); i.e., to perform the scale analysis, instead of the variables, the maximum values which they assume inside the boundary layer are used. In this case, only the parameter x will be left as really variable in order to obtain results in the function of it (in particular, the layer thickness δ will be obtained as a function of x), and the velocity v. It is possible, therefore, to write:

$$U_\infty^2/x \text{ instead of } u\,\partial u/\partial x \qquad (7)$$

$$vU_\infty/\delta \text{ instead of } v\,\partial u/\partial y \qquad (8)$$

$$\nu U_\infty/\delta^2 \text{ instead of } \nu\,\partial^2 u/\partial y^2 \qquad (9)$$

The following formal notation is written:

$$(U_\infty^2/x,\ vU_\infty/\delta) \sim \nu\,U_\infty/\delta^2 \qquad (10)$$

with this meaning: the first member of Equation 5 has the same order of magnitude of the second. Equation 6 establishes that:

$$U_\infty/x \sim v/\delta \quad (11)$$

i.e., that (U_∞/x) has the same order of magnitude as (v/δ); it follows also that the two inertial terms of the first member of Equation 5 have the same order of magnitude and, therefore, both of them have the same order of magnitude with respect to the second member. It is, therefore:

$$v\, U_\infty/\delta \sim \nu\, U_\infty/\delta^2; \quad U_\infty^2/x \sim \nu\, U_\infty/\delta^2 \quad (12)$$

From the second relation, it is possible to obtain:

$$\delta \sim (\nu x/U_\infty)^{1/2} \quad \text{or} \quad \delta/x \sim Re_x^{-1/2} \quad (13)$$

Such a relation represents the equilibrium inside the boundary layer between the viscous and the inertial forces. Only with simple considerations on the orders of magnitude of the main parameters (or of groups of them), has it been possible to obtain functional links among the various physical quantities, within proportionality constants between 0 and 1. Such constants may be exactly determined using analytical or numerical methods for the solution of conservation equations.

Finally, the rules to which the scale analysis obeys are the following:

1. The spatial extension of the region where the analysis is performed must be defined
2. In each equation representing an equilibrium between or among members, the order of magnitude of each term must be evaluated, eventually neglecting it
3. All the physical quantities must be expressed with their maximum value, with the exception of the independent variables. For better comprehension, it is advisable to consult Section III: Applications.

D. Integral Method

This is an approximate method for the solution of the conservation equations, developed by Pohlhausen and Von Kármán at the beginning of this century; it is generally used in applications only for the determination of heat transfer between the wall and the fluid. The integral method consists in the integration of the differential conservation equations, making use of reasonable temperature and velocity profiles, *a priori* hypothesized. The procedure is the following (see Figure 3):

1. The profiles must be expressed in function of the flow direction and of δ and δ_T. δ and δ_T are the unknown quantities to be determined through equations integration.
2. The equations are solved in the unknowns δ and δ_T, allowing the evaluation of the fluidodynamic and of the thermal layer thicknesses.
3. δ and δ_T, substituted in the velocity and temperature profiles *a priori* hypothesized, allow the determination of the velocity and temperature gradients.
4. Heat flux q'' may then be determined through Equation 4.

Iterations may be necessary.

The arbitrary nature of the velocity and temperature profiles' introduction makes this procedure approximate. Thanks to the small thicknesses of the boundary layers, however, the mistake associated with the assumption of an approximate profile is generally very small.

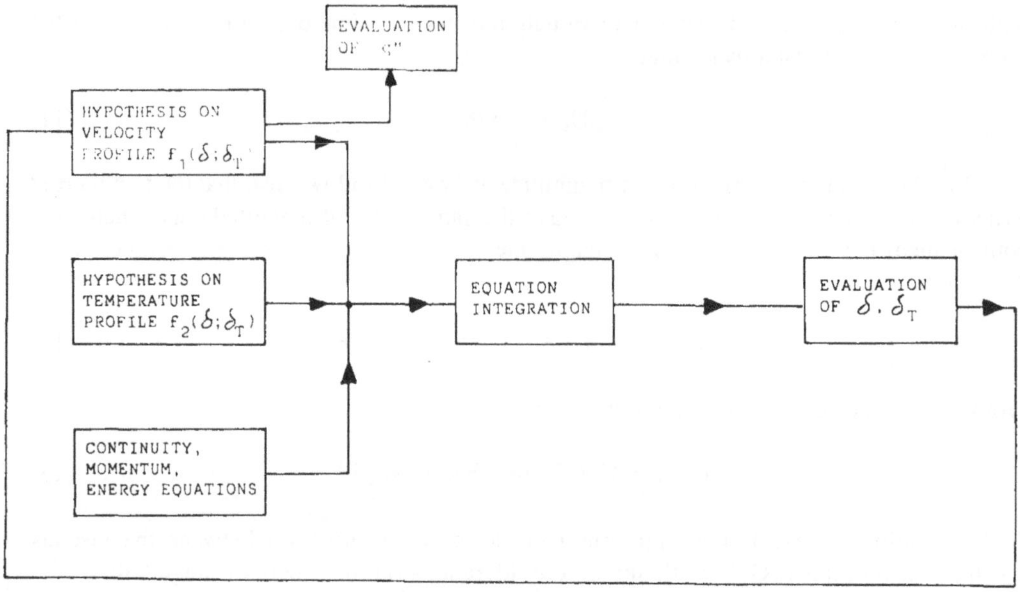

FIGURE 3. Integral method flow diagram.

III. APPLICATIONS

Within the limits of the present text, the situation of a vertical plane slab is given special consideration because of its relevance in applications.

For the solution of problems pertaining to different configurations (cavities, porous media, etc.), some miscellaneous correlations are reported in Section B (see also References 8 to 16).

A. Vertical Slab

1. Solving Equations

With reference to Figure 2, a plane slab of finite height H, submerged in a semi-infinite fluid at temperature T_∞ is considered. The boundary conditions are

$$\left.\begin{aligned} u = v = 0 \\ T = T_w \end{aligned}\right\} x = 0 \qquad (14a)$$

$$\left.\begin{aligned} v = 0 \\ T = T_\infty \end{aligned}\right\} x \rightarrow \infty \qquad (14b)$$

where the second condition is a consequence of the nature of the external flow current closure, regarded as of infinite length (descending fluid velocity = 0).

With reference to Section II, the following four considerations may be made:

- The flow is stationary, so that the term ($\partial v/\partial t$) of the momentum equation in y direction is null (see Chapter 2, Equation 9b).
- The analysis is performed with reference to the region inside the kinematic boundary

layer; it is very thin with respect to the height H of the slab, so that among the inertial force terms it is

$$\partial^2 v/\partial y^2 \ll \partial^2 v/\partial x^2 \qquad (15)$$

In fact, it is (Section II.C):

$$x \sim \delta; \quad y \sim H$$

- In general, the pressure depends both on x and y, so that its total differential is

$$dp = (\partial p/\partial x)\, dx + (\partial p/\partial y)\, dy \qquad (16)$$

from which:

$$dp/dy = (\partial p/\partial x)\,(dx/dy) + \partial p/\partial y \qquad (17)$$

Considering the orders of magnitude, it may be assumed that ($\partial p/\partial x \simeq 0$), from which:

$$dp/dy = dp_\infty/dy \simeq \partial p/\partial y \qquad (18)$$

where p_∞ is the value at a distance ∞ from the wall. This term is the hydrostatic pressure gradient and may be expressed as $\partial p/\partial y = -\rho_\infty g$, where ρ_∞ is the value of ρ at distance ∞ from the wall. The term (see Chapter 2, Equation 9b) $-\partial p/\partial y - \rho g$, therefore, becomes $g(\rho_\infty - \rho)$. In general, the expression of density in function of temperature may be represented through a series, considering only the first two terms (limited temperature variation). As a consequence, if β is the volumetrical expansion coefficient at constant pressure:

$$\rho \simeq \rho_\infty[1 - \beta(T - T_\infty)]; \quad \beta = -(1/\rho)\,(\partial \rho/\partial T)_{p=\text{const}} \qquad (19)$$

- Density gradients are, in general, small, so that the fluid may be considered incompressible. The only term in which variations of density must be considered is the volume forces term. As a consequence, the momentum, energy, and continuity equations become, in the hypothesis of constant physical properties:

$$u\frac{\partial v}{\partial x} + v\frac{\partial v}{\partial y} = \nu\frac{\partial^2 v}{\partial x^2} + \beta g(T - T_\infty) \quad \text{(momentum equation, y-axis)} \qquad (20)$$

$$u\frac{\partial T}{\partial x} + v\frac{\partial T}{\partial y} = \alpha\frac{\partial^2 T}{\partial x^2} \quad \text{(energy equation)} \qquad (21)$$

$$\frac{\partial u}{\partial x} + \frac{\partial v}{\partial y} = 0 \quad \text{(continuity equation)} \qquad (22)$$

Such equations, in the unknown quantities u, v, and T completely determine the problem of defining the velocity and temperature profiles inside the boundary layers.

2. Scale Analysis

The initial information on the thermofluidodynamic problem is obtained through a scale analysis applied to Equations 20 to 22.[1] With reference to the region adjacent to the slab, which is thermally affected by its presence (δ_T is its thickness), it may be assumed in stationary conditions that through this fluid layer heat is transmitted only by conduction. It will be integrally removed by the fluid in motion and there will be an equilibrium between the convection effect of the fluid and the conduction effect through the boundary layer on the slab. This may be expressed through the energy equation, where the various orders of magnitude are

$$x \sim \delta_T; \quad y \sim H; \quad (T - T_\infty) \sim (T_w - T_\infty)$$

If Equation 21 is written in terms of scale, it is

$$[u(T_w - T_\infty)/\delta_T, \ v(T_w - T_\infty)/H] \sim \alpha(T_w - T_\infty)/\delta_T^2 \tag{23}$$

In the first two terms the convection effect may be recognized, while the third term is the conduction term.

Due to the continuity equation, which establishes the following relationship:

$$u/\delta_T \sim v/H \tag{24}$$

we know that the two convection terms have the same order of magnitude with respect to the third term. The following scale relation is obtained:

$$v \ \Delta T/H \sim \alpha \Delta T/\delta_T^2 = \rightarrow v \sim \alpha \ H/\delta_T^2 \tag{25}$$

If Equation 20 is considered, it represents the equilibrium among three forces: inertial, friction, and buoyancy. The aim is to establish the conditions under which inside the boundary layers there is an equilibrium of the kind inertial-buoyancy, or friction-buoyancy, as the buoyancy effect is always present (the analysis is performed inside the region where density gradients occur in x direction).

The inertia terms are (following Equation 20) uv/δ_T, v^2/H, and have the same order of magnitude; the friction term is $\nu v/\delta_T^2$; and the buoyancy term is $g \ \beta \Delta T$. If all these terms are divided by $g \ \beta \Delta T$, if v is substituted by means of Equation 25 and if Rayleigh and Prandtl numbers are introduced (Chapter 1):

$$Ra_H = g \ \beta \ \Delta \ T \ H^3/(\alpha \ \nu); \quad Pr = \nu/\alpha \tag{26}$$

it is obtained as:

$$\bullet \ \text{Inertia} \rightarrow (H/\delta_T)^4 \ Ra_H^{-1} \ Pr^{-1} \tag{27}$$

$$\bullet \ \text{Friction} \rightarrow (H/\delta_T)^4 \ Ra_H^{-1} \tag{28}$$

$$\bullet \ \text{Buoyancy} \rightarrow 1 \tag{29}$$

From these relations, it may be noted that in order to compare the importance of the inertia and the friction terms, a characteristic of the fluid is fundamental: the Prandtl number. This is the reason it is necessary to distinguish between situations with high or low values of the Prandtl number.

a. High Values of Pr: Pr ≫ 1

From Equations 27 and 28, the inertia term may be neglected with respect to the friction

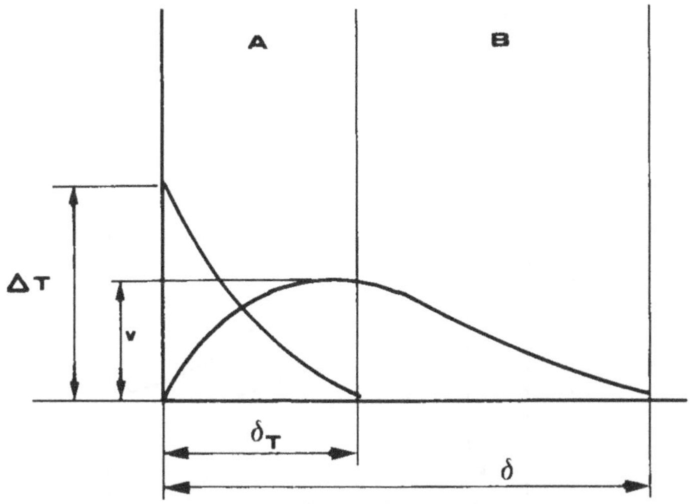

FIGURE 4. Velocity and temperature profiles (Pr ≫ 1). (A) Driving heated layer, friction ~ buoyancy; (B) driven unheated layer, friction ~ inertia.

term. In this situation, inside δ_T there will be an equilibrium of the friction-buoyancy kind, so that from Equations 28 and 29:

$$(H/\delta_T)^4 \, Ra_H^{-1} \sim 1 \rightarrow \delta_T \sim H \, Ra_H^{-1/4} \tag{30}$$

so that, using Equation 25:

$$v \sim (\alpha/H) \, Ra_H^{1/2} \tag{31}$$

In order to evaluate the heat transfer coefficient, in the hypothesis of a laminar flow and of a linear temperature field inside δ_T:

$$h = k/\delta_T \tag{32}$$

so that it may be assumed that the order of magnitude of h is

$$h \sim k/\delta_T \tag{33}$$

If a Nusselt number is introduced as

$$Nu_H = h \, H/K \tag{34}$$

from Equations 30 and 33:

$$Nu_H \sim Ra_H^{1/4} \tag{35}$$

For what regards velocity v, some considerations are necessary, also considering Figure 4, which represents a qualitative trend of velocity and temperature profiles. The velocity is different from 0 outside the thermal boundary layer, where the buoyancy effects are null. In this zone, a flow occurs because of the viscous drag due to the fluid in motion inside δ_T. In this region, the equilibrium of forces is of the friction-inertia kind, so that inside this boundary layer (called the velocity layer δ), the following relation may be assumed:

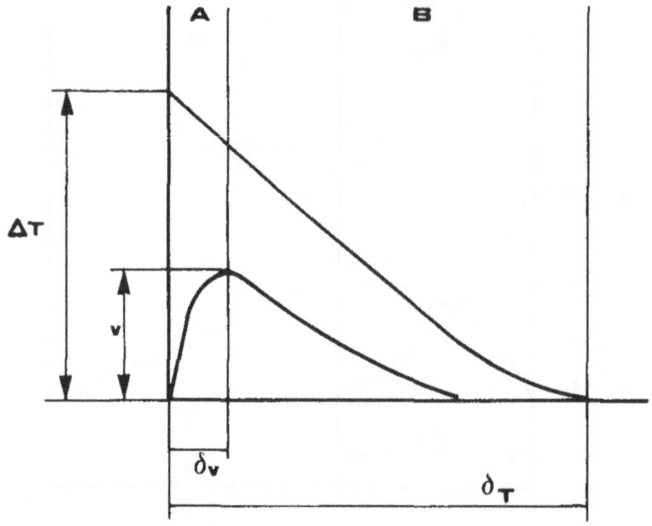

FIGURE 5. Velocity and temperature profiles ($Pr \gg 1$). (A) Friction ~ buoyancy; (B) inertia ~ buoyancy.

$$v^2/H \sim \nu\ v/\delta^2 \tag{36}$$

With the introduction of the Rayleigh and Prandtl numbers and considering Equation 31, the kinematic (velocity) boundary layer is

$$\delta \sim H\ Ra_H^{-1/4}\ Pr^{1/2} \tag{37}$$

If the ratio δ/δ_T is considered, with high values of the Prandtl number (i.e., for very viscous fluids with a low conductivity), the velocity boundary layer will be always thicker than the thermal boundary layer (see Equation 30):

$$\delta/\delta_T \sim Pr^{1/2} > 1 \tag{38}$$

b. Low Values of Pr: Pr ≪ 1

From Equations 24, 29, and 31, the following relations may be obtained:

$$\delta_T \sim H(Ra_H\ Pr)^{-1/4} \tag{39}$$

$$v \sim (\alpha/H)\ (Ra_H\ Pr)^{1/2} \tag{40}$$

$$Nu \sim (Ra_H\ Pr)^{1/4} \tag{41}$$

which are originated by a dynamic equilibrium of the inertia-buoyancy kind, valid for fluids with low viscosity and high thermal conductivity.

It is seen (Figure 5) that due to the low viscosity, the flow exists only where gradients of temperature exist, while it is null outside δ_T. With considerations analogous to those previously made, the thickness of the velocity boundary layer δ_v may be obtained, representing the layer thickness corresponding to the peak of velocity:

$$\delta_v \sim H(Ra_H/Pr)^{-1/4} = H\ Gr_H^{-1/4} \tag{42}$$

being $Gz_H = Ra_H/Pz$.

Table 1
SCALE ANALYSIS RESULTS

Prandtl no. range	Thermal boundary layer thickness	Wall jet velocity profile: Distance from wall to velocity peak	Wall jet velocity profile: Thickness of wall jet	Wall jet velocity profile: Velocity scale	Nusselt no. $Nu = \frac{hH}{k}$
$Pr > 1$	$H\,Ra_H^{-1/4}$	$H\,Ra_H^{-1/4}$	$Pr^{1/2}(H\,Ra_H^{-1/4})$	$\frac{\alpha}{H}\,Ra_H^{1/2}$	$Ra_H^{1/4}$
$Pr < 1$	$Pr^{-1/4}(H\,Ra_H^{-1/4})$	$Pr^{1/4}(H\,Ra_H^{-1/4})$	$Pr^{-1/4}(H\,Ra_H^{-1/4})$	$\frac{\alpha}{H}\,(Pr\,Ra_H)^{1/2}$	$(Pr\,Ra_H)^{1/4}$

From Equations 39 and 42:

$$\delta_v/\delta_T \sim Pr^{1/2} < 1 \tag{43}$$

The results are grouped in Table 1.[1]

3. Integral Method

The integration of conservation equations is performed from the plane $x = 0$ to the plane $x = X$, in correspondence to which the flow is null and the temperature is T_∞. The integral equations of the boundary layer for momentum and energy are obtained from Equation 20 and 29 and are[1]

$$\frac{d}{dy}\int_0^X v^2\,dx = -\nu\left(\frac{\partial v}{\partial x}\right)_{x=0} + g\,\beta\int_0^X (T - T_\infty)\,dx \tag{44}$$

$$\frac{d}{dy}\int_0^X v(T_\infty - T)\,dx = \alpha\left(\frac{\partial T}{\partial x}\right)_{x=0} \tag{45}$$

In order to solve them, it is necessary to hypothesize a priori velocity and temperature profiles (see Section II.D). Furthermore, it is necessary to separately treat the situation of $Pr \gg 1$ and that of $Pr \ll 1$.

a. High Values of Pr: Pr ≫ 1

In this situation, suitable velocity and temperature profiles are[1]

$$T - T_\infty = \Delta\,T\,\exp(-x/\delta_T) \tag{46}$$

$$v = V\,\exp(-x/\delta)\,[1 - \exp(-x/\delta_T)] \tag{47}$$

where V, δ_T, and δ are unknown functions of y.

There are, therefore, three unknown quantities with two integral equations and a third condition is necessary. A method is that proposed by Squire, who poses $\delta_T = \delta$, solving the equations. In general, it is $\delta_T \neq \delta$, so that further considerations are necessary; one of them concerns the low value of inertia forces in the sublayer near the wall (in $0 < x < 0^+$, fluid is motionless).

If to the ratio δ/δ_T symbol q is assigned, the dependence of such ratio on Pr leads to the following relationships:

$$Pr = (5/6)\,q^2(q + 0.5)/(q + 2) \tag{48}$$

$$Nu = \frac{q''}{T_w - T_\infty}\frac{y}{k} = \left[\frac{3}{8}\frac{q^3}{(q+1)(q+0.5)(q+2)}\right]^{1/4} Ra_y^{1/4} \tag{49}$$

which solve the problem of the evaluation of heat transferable from the slab to the fluid. If $Pr \rightarrow \infty$, Equation 49 reduces to:

$$Nu = 0.783\ Ra_y^{1/4} \tag{50}$$

b. Low Values of Pr: Pr ≪ 1

The temperature profile used is the same as in the preceding situation, while for the velocity it is assumed that:

$$v = V_1 \exp(-x/\delta_T)\ [1 - \exp(-x/\delta_v)] \tag{51}$$

where V_1, δ_T, and δ_v are unknown functions of y. If Equations 44 and 45 are considered and assuming as null inertia forces on the wall sublayer, if δ_v/δ_T is defined as q_1[1]:

$$Pr = (5/3)\ [q_1/(1 + q_1)]^2 \tag{52}$$

$$Nu = (3/8)^{0.25}\ [q_1/(2\ q_1 + 1)]^{1/2} \cdot Ra_y^{1/4} \tag{53}$$

which, if $Pr \rightarrow 0$, lead to:

$$\delta_v/\delta_T = \left(\frac{3}{5}Pr\right)^{0.5} \tag{54}$$

$$Nu = 0.689(Pr\ Ra_y)^{1/4} \tag{55}$$

B. Miscellaneous Correlations

•Spherical bodies: If D is the sphere diameter[17]:

$$\overline{Nu_D} = 2 + 0.45(Gr_D \cdot Pr)^{1/4} \tag{56}$$

- Vertical enclosures with air: If δ is the distance between two vertical slabs, with height H and with a difference of temperature ΔT[18]:

$$\overline{Nu_\delta} = \begin{cases} 0.18\ Gr_\delta^{1/4}\ (H/\delta)^{-1/9} & 20{,}000 < Gr_\delta < 200{,}000 \\ 0.065\ Gr_\delta^{1/3}\ (H/\delta)^{-1/9} & 200{,}000 < Gr_\delta < 11 \cdot 10^6 \end{cases} \tag{57}$$

- Horizontal enclosures with air: If δ is the distance between two horizontal slabs[8]:

$$\overline{Nu_\delta} = \begin{cases} 0.195\ Gr_\delta^{1/4} & 1 \cdot 10^4 < Gr_\delta < 4 \cdot 10^5 \\ 0.068\ Gr_\delta^{1/3} & 4 \cdot 10^5 < Gr_\delta \end{cases} \tag{58}$$

REFERENCES

1. **Bejan, A.,** *Convection Heat Transfer,* John Wiley & Sons, New York, 1984.
2. **Rohsenow, W. M. and Choi, H. Y.,** *Mass and Momentum Transfer,* Prentice-Hall, Englewood Cliffs, N.J., 1961.
3. **Sparrow, E. M. and Gregg, J. L.,** Laminar free convection from a vertical plate with uniform surface heat flux, *Trans. ASME,* 78, 1956.
4. **Schlichting, H.,** *Boundary Layer Theory,* McGraw-Hill, New York, 1979.
5. **Eckert, E. R. G. and Drae, R. M.,** *Analysis of Heat and Mass Transfer,* McGraw-Hill, New York, 1972.
6. **Le Fevre, E. J.,** Laminar free convection from a vertical plane surface, 9th Int. Congr. Applied Mechanics, Brussels, 1956.
7. **Yang, K. T.,** Possible similarity solutions for laminar free convection on vertical plates and cylinders, *J. Appl. Mech.,* 27, 1960.
8. **Anderson, R. and Bejan, A.,** Natural convection of both sides of a vertical wall separating fluids at different temperature, *J. Heat Transfer,* 102, 1980.
9. **Patterson, J. and Imberger, I.,** Unsteady natural convection in a rectangular cavity, *J. Fluid Mech.,* 100, 1980.
10. **Gill, A. E.,** The boundary layer regime for convection in a rectangular cavity, *J. Fluid Mech.,* 26, 1966.
11. **Graebel, W. P.,** The influence of Prandtl number on free convection in a rectangular cavity, *Int. J. Heat Mass Transfer,* 24, 1981.
12. **Bejan, A., Al-Homoud, A. A., and Imberger, I.,** Experimental study of high Rayleigh number convection in a horizontal cavity with different end temperatures, *J. Fluid Mech.,* 109, 1981.
13. **Batchelor, G. K.,** Heat transfer by free convection across a closed cavity between vertical boundaries at different temperature, *Q. Appl. Math.,* 12, 1954.
14. **Eckert, E. R. G. and Carlson, W. O.,** Natural convection in an air layer enclosed between two vertical plates with different temperatures, *Int. J. Heat Mass Transfer,* 2, 1961.
15. **Bejan, A. and Anderson, R.,** Heat transfer across a vertical impermeable partition unbebbed in a porous medium, *Int. J. Heat Mass Transfer,* 24, 1981.
16. **Bejan, A. and Anderson, R.,** Natural convection at the interface between a vertical porous layer and an open space, *J. Heat Transfer,* 105, 1983.
17. **Yuge, T.,** Experimental on heat transfer from spheres including combined natural and forced convection, *J. Heat Transfer,* 82, 1960.
18. **Jackob, M.,** *Heat Transfer,* John Wiley & Sons, New York, 1949.
19. **Kakaç, S., Aung, W., and Viskanta, R.,** *Natural Convection — Fundamentals and Applications,* Hemisphere, New York, 1985.

Chapter 6

CRITICAL FLOWS

I. INTRODUCTION

The flow rate through a channel or an orifice increases with the increase of the pressure difference Δp between the two sections (or volumes), one upstream (feeding pressure vessel) and the other downstream (outlet receiver). There is a value of Δp in correspondence to which the flow rate reaches a maximum value (critical flow rate) and stabilizes, even if Δp increases (with a constant upstream pressure and a downstream pressure decrease). For a given channel with a given inlet pressure, the maximum static pressure at its outlet section, leading to the critical flow rate, is called "critical pressure". Its value depends on the channel geometry and on the physical properties of the fluid upstream of the channel (inlet conditions).

Considering the channel outlet section, if a decrease in the outlet receiver pressure does not determine an increase in flow rate (pressure lower than the critical one), it means that the decreasing-pressure effect cannot be transmitted upstream in the flowing fluid. That is, the fluid velocity has the same value of the propagation velocity of pressure waves in the fluid itself. This velocity is called "sonic velocity"; in critical flows, the fluid (which accelerates along the discharging channel) reaches, at the outlet section, a velocity that equals its sonic velocity, which is a function of the critical pressure and of other physical conditions at the outlet section (as, for instance, void fraction and phase distribution). In this sense, the determination of the critical flow conditions is relatively easy in single-phase flows, whereas in multiphase flows, it becomes more difficult owing to the interactions among the phases, which play a determinant role.

II. SINGLE-PHASE CRITICAL FLOWS

A. Acoustic Waves and Sonic Velocity

In a single-phase flow, the acoustic waves (pressure oscillations, small with respect to the static pressure) may be described through the following simplified form of the momentum equation:

$$\partial v/\partial t + v\, \partial v/\partial x = -(1/\rho)\,(\partial p/\partial x) \tag{1}$$

where the spatial derivative $\partial v/\partial x$ is usually very small and may be neglected:

$$\partial v/\partial t \simeq -(1/\rho)\, \partial p/\partial x \tag{2}$$

An acoustic wave travels at the sonic velocity "a": in a time Δt, a distance $\pm$ a Δt (being the sign + [or −] corresponding to the same [or contrary] direction of the fluid velocity v) is covered.

If θ is defined as $\theta = x \pm at$, after substitution:

$$\pm\, a\, \partial v/\partial \theta = -(1/\rho)\,(\partial p/\partial \theta) \tag{3}$$

After integration:

$$\Delta p/\Delta v = \pm\, \rho a \tag{4}$$

Pressure waves (both of compression or rarefaction) propagate along the channel (piping) and whenever they meet rigid obstacles, reflect backward. If they meet an open outlet or a big plenum, the reflection is characterized by waves of the same amplitude Δp with an opposite sign. If the channel has an abrupt variation in the cross flow area, from A_1 to A_2, part of the incident pressure wave Δp_{inc} is transmitted (Δp_{tr}) and part is reflected backward (Δp_{ref}):

$$\Delta p_{tr}/\Delta p_{inc} = 2\,A_1/(A_1 + A_2) \tag{5}$$

$$\Delta p_{ref}/\Delta p_{inc} = (A_1 - A_2)/(A_1 + A_2) \tag{6}$$

Wall friction and viscous dissipations reduce, with time and along a piping, the wave amplitude:

$$\Delta p = \Delta p_o \exp(-f\ v\ t/d) \tag{7}$$

Equation 7 allows the evaluation of the wave propagation time constant; if, for instance, the wave propagates in a channel with a diameter $d = 0.3$ m, a wall friction factor $f = 0.005$, and with a fluid velocity $v = 6$ m/sec, the time constant d/fv equals 10 s.

In homogeneous and isotropic fluids (gases and liquids), the sonic velocity depends mainly on density and increases coherently. In gases and vapors, its value is much lower than in liquids; for this reason, critical flows occur more frequently with the previous ones.

In dishomogeneous and anisotropic fluids (two-phase mixtures), the sonic velocity is strongly influenced by the void fraction and the flow regime in the critical section (outlet section of the discharging channel).

As is well known from classic thermodynamics, the sonic velocity for perfect gases is given by:

$$a = \sqrt{(\partial p/\partial \rho)_s} \tag{8}$$

This expression may be applied, with sufficient accuracy, also to superheated vapors. For two-phase, single-component mixtures there is currently no generally accepted expression for the sonic velocity. It is possible, indeed, to formulate two quite opposite, simplifying hypotheses, when a small pressure pulse propagates:

1. The thermodynamic equilibrium is preserved, with instantaneous mass transfer between the two phases
2. No mass transfer happens (frozen state), with independent, isentropic transformations of each phase

When the mixture may be considered as homogeneous (bubbly flow, small α; dispersed flow, $\alpha \simeq 1$), it is possible to introduce an average mixture density:

$$\bar{\rho} = \alpha\,\rho_g + (1 - \alpha)\,\rho_l$$

allowing the evaluation of the sonic velocity in "frozen state":

$$a_{tp}^2 = \left[\frac{\alpha\,\bar{\rho}}{\rho_g\,a_g^2} + \frac{(1 - \alpha)\,\bar{\rho}}{\rho_l\,a_l^2}\right]^{-1} \tag{9}$$

where a_g, a_l, and a_{tp} are, respectively, the sonic velocities in the vapor, in the liquid, and in the two-phase mixture. At low pressure, $a_l >> a_g$, so that:

$$a_{tp}^2/a_g^2 = [\alpha^2 + \alpha(1 - \alpha)\ \rho_l/\rho_g]^{-1} \tag{10}$$

At intermediate values of the void fraction α, it is necessary to consider the different velocities of the two phases and more complex expressions may be formulated. For instance, with an annular or stratified flow and with the hypothesis of "frozen state", it is suggested[16]:

$$a_{tp}^2 = \frac{\{[(1 - X)\ \rho_g + X\ \rho_l]^2 + X(1 - X)\ (\rho_l - \rho_g)^2\}}{\left[\dfrac{X \cdot \rho_l^2}{a_g^2}\right] + \left[\dfrac{(1 - X)\ \rho_g^2}{a_l^2}\right]} \tag{11}$$

B. Critical Flow Rate

If a single-phase, adiabatic flow through an ideal nozzle is considered, the conservation of enthalpy from the inlet (stagnation conditions) to the outlet section may be expressed as:

$$i_o = i + v^2/2 = i + G^2/(2\ \rho^2) \tag{12}$$

From the first and second law of thermodynamics:

$$di = du + pd\left(\frac{1}{\rho}\right) + \left(\frac{1}{\rho}\right) dp = dq - pd\left(\frac{1}{\rho}\right) + pd\left(\frac{1}{\rho}\right) + \frac{1}{\rho}\ dp =$$

$$= T\ ds + \left(\frac{1}{\rho}\right) dp$$

In adiabatic hypothesis:

$$T\ ds = 0; \quad di = (1/\rho)\ dp$$

From Equation 12, the mass velocity may be obtained:

$$G = \rho[2(i_o - i)]^{1/2} = \rho\left(2 \int_p^{p_o} \frac{dp}{\rho}\right)^{1/2} \tag{13}$$

which, with incompressible fluids (being $\rho = \rho_l =$ constant), gives:

$$G = [2\ \rho_l\ p_o(1 - p/p_o)]^{1/2} \tag{14}$$

The maximum (critical) value of G is obtained substituting p_{cr}/p_o to p/p_o, p_{cr} being the critical static pressure in the receiver volume.

With the perfect gases, the adiabatic transformation (expansion) is

$$p(1/\rho)^\gamma = p_o(1/\rho_o)^\gamma \tag{15}$$

Equation 15 may be used to substitute $1/\rho$ in Equation 13, leading to:

$$G = \left\{2\frac{\gamma}{\gamma - 1}\ p_o\ \rho_o\left[\left(\frac{p}{p_o}\right)^{\frac{2}{\gamma}} - \left(\frac{p}{p_o}\right)^{\frac{\gamma+1}{\gamma}}\right]\right\}^{1/2} \tag{16}$$

The critical condition is

$$d\,G/dp = 0 \tag{17}$$

which gives:

$$p_{cr}/p_o = [2/(\gamma + 1)]^{\gamma/(\gamma - 1)} \tag{18}$$

Therefore, a perfect-gas critical mass velocity is

$$G_{cr} = \left[\gamma\, p_o\, \rho_o \left(\frac{2}{\gamma + 1}\right)^{\left[\frac{\gamma+1}{\gamma-1}\right]}\right]^{1/2} \tag{19}$$

With real gases and with superheated steam (at least at low pressure), it is possible to use, instead of Equation 15, the politropic expansion relationship:

$$p(1/\rho)^n = \text{const} \tag{20}$$

with the politropic exponent $n = 1.3$. The mass velocity may be expressed as:

$$G = \left\{\frac{2\,n}{n - 1}\, p_o\, \rho_o \cdot \left[\left(\frac{p}{p_o}\right)^{\frac{2}{n}} - \left(\frac{p}{p_o}\right)^{\frac{n+1}{n}}\right]\right\}^{(1/2)} \tag{21}$$

the critical pressure ratio is

$$p_{cr}/p_o = [2/(n + 1)]^{n/(n - 1)} \tag{22}$$

the critical mass velocity is

$$G_{cr} = \left[n\, p_o\, \rho_o \left(\frac{2}{n + 1}\right)^{(n + 1)/(n - 1)}\right]^{1/2} \tag{23}$$

For biatomic gases and for superheated steam, respectively, $n = 1.41$, $p_{cr}/p_o = 0.528$ and $n = 1.3$, and $p_{cr}/p_o = 0.546$.

Figure 1 shows the trends of $G/(2\, p_o\, \rho_o)^{1/2}$ vs. p/p_o for a liquid and a gas, the static pressure p being measured in the outlet section of the nozzle.

III. TWO-PHASE CRITICAL FLOWS

The critical flow through a channel of an initially subcooled or saturated liquid (at stagnation conditions) is considered. Depressurization causes the onset of vaporization and of a two-phase flow, with mechanical and thermal interactions between the phases. Several models have been developed to interpret the phenomenon; Figure 2 synthesizes their classification and connections.

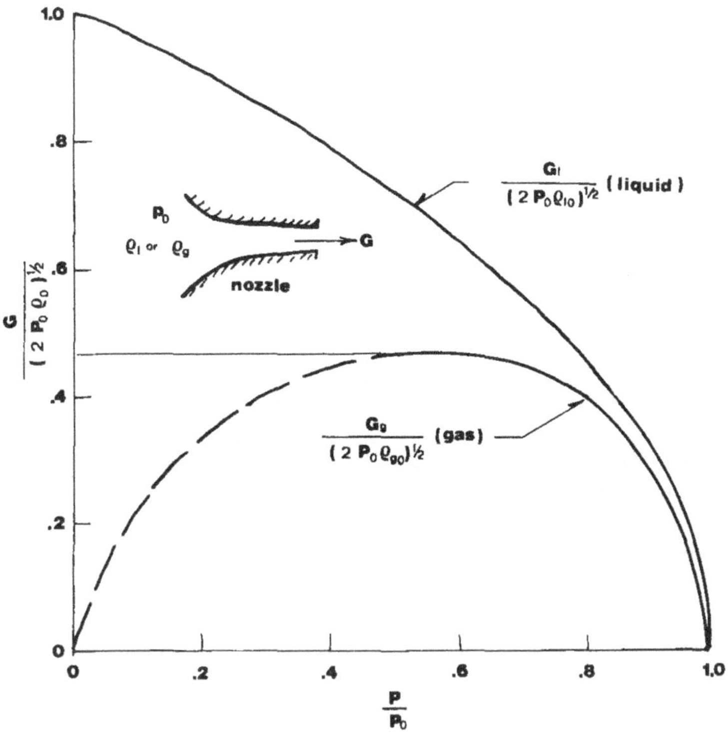

FIGURE 1. Mass velocity as a function of downstream pressure.

A. Thermodynamic-Equilibrium Models

1. Homogeneous Model[15]

This model hypothesizes, during expansion, a thermodynamic equilibrium and equal velocities for the two phases; in these conditions:

$$G^2 = -\left[X \cdot \frac{d\left(\frac{1}{\rho_g}\right)}{dp} + \left(\frac{1}{\rho_g} - \frac{1}{\rho_l}\right)\frac{dX}{dp} + (1 - X) \cdot \frac{d\left(\frac{1}{\rho_l}\right)}{dp}\right]^{-1} \tag{24}$$

If the liquid-phase compressibility is neglected and quality is expressed as a function of entropy:

$$G^2 = -\left[X \cdot \frac{d\left(\frac{1}{\rho_g}\right)}{dp} - \frac{1}{\rho_g}\left(\frac{X}{s_g - s_l}\frac{d\,s_g}{dp} + \frac{1 - X}{s_g - s_l}\frac{d\,s_l}{dp}\right)\right]^{-1} \tag{25}$$

where the variables are computed at the outlet conditions.

Alternatively, the mass velocity may be obtained employing the energy equation in adiabatic conditions (enthalpies computed at p_o):

$$G = \frac{\sqrt{2[i_o - (1 - X)\,i_l - Xi_g]}}{[(1 - X)/\rho_l + X/\rho_g]} \tag{26}$$

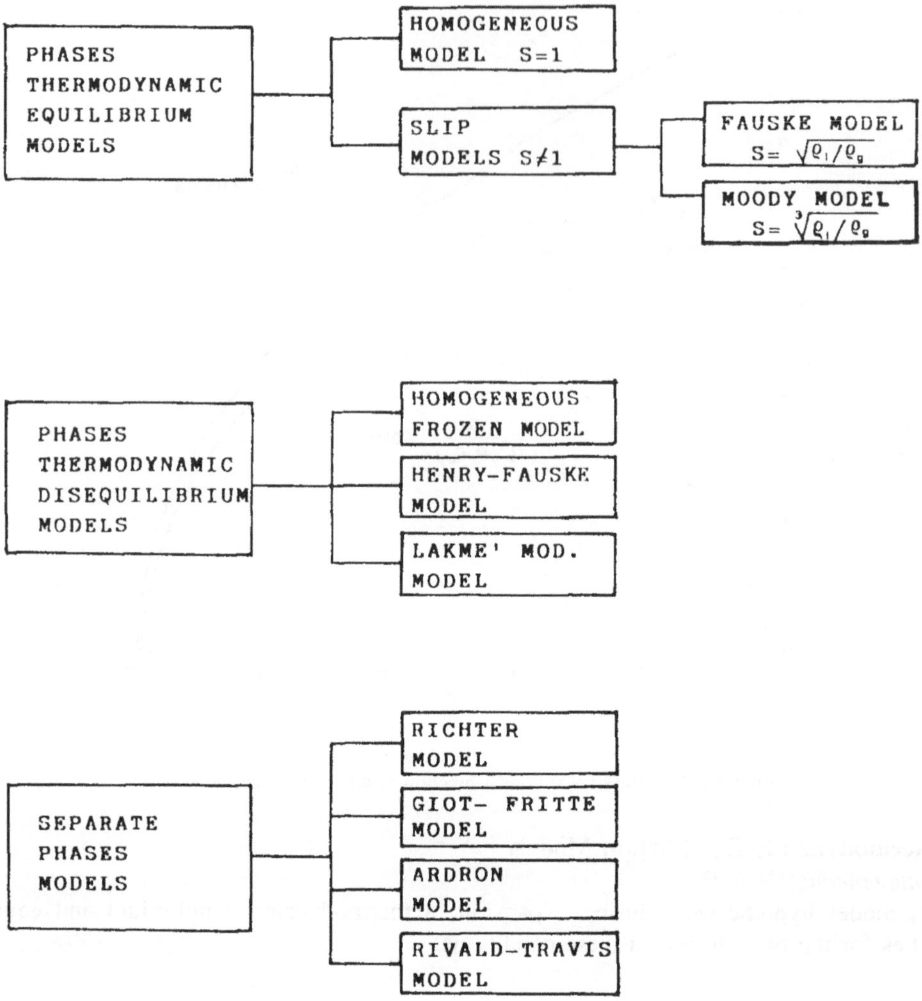

FIGURE 2. Two-phase critical flow models.

To compute the critical value of the flow rate from the above expressions, it is sufficient to impose the critical condition:

$$(\partial G/\partial p)_{G=G_{cr}} = 0 \tag{27}$$

2. *Slip Models*

Some authors[2,3] suggest the admission of a slip between the mean velocities of the two phases, with a closer description of the physical behavior.

a. Fauske Model[2]

This model rests on the following hypotheses: no heat and mass transfer between the phases; no friction; thermodynamic equilibrium; slip between the phases; and annular flow with liquid on the wall. Imposing in the critical section $\partial G/\partial p = 0$ and slip such as to minimize the specific momentum, slip ratio is

$$S^* = (\rho_l/\rho_g)^{1/2} \tag{28}$$

and the critical mass velocity is obtained:

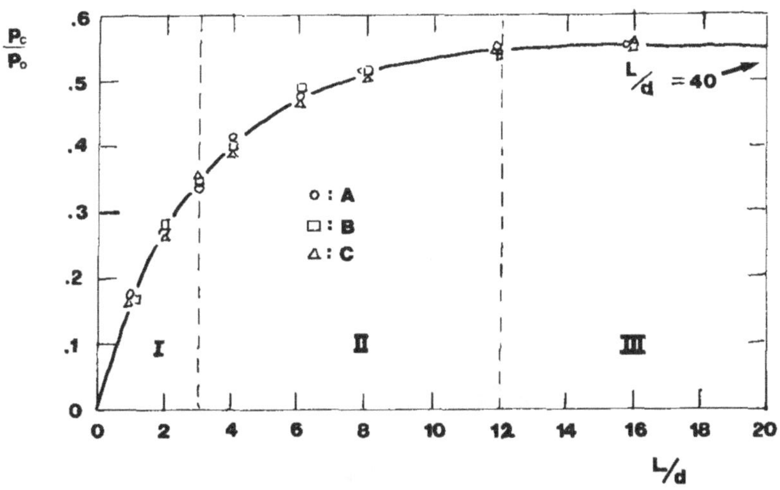

FIGURE 3. Critical pressure ratio vs. L/d. (A) Low p; (B) intermediate p; (C) high p.

$$G_{cr}^2 = \frac{-S^*}{[(1 - X + S^*X)X]\dfrac{d\left(\frac{1}{\rho_g}\right)}{dp} + \left[\dfrac{1}{\rho_g}(1 + 2\,S^*X - 2\,X) + \dfrac{1}{\rho_l}(2S^*X - 2S^* - 2XS^{*2} + S^{*2})\right]\dfrac{dX}{dp}} \tag{29}$$

Figure 3 provides the critical pressure ratio p_{cr}/p_o vs. the length-to-diameter ratio of the discharging channel, showing that only for L/d >12 this ratio stabilizes. It has been found[4] that the critical ratio p_{cr}/p_o, at a given stagnation condition, depends on L/d only for L/d <20, while a clear influence is exerted by the inlet subcooling of the liquid (Figure 4). This model agrees very well with experimental data for L/d >12 and p_{cr} <30 bar.

b. Moody Model[3]

In this model, which refers to annular flow with liquid on the wall, the slip ratio in the critical section (outlet) is obtained with the hypothesis that it minimizes the mixture's specific kinetic energy:

$$S^* = (\rho_l/\rho_g)^{1/3} \tag{30}$$

In short channels, mass velocity may be expressed as a function of s (enthropy) and p (enthalpies and entropies evaluated at p_o):

$$G(S^*, p) = \left\{ \frac{2\left[i_o - i_l - \lambda\left(\dfrac{s_o - s_l}{s_g - s_l}\right)\right]}{\left[S^*\dfrac{(s_g - s_l)}{\rho_l(s_g - s_l)} + \dfrac{(s_o - s_l)}{(s_g - s_l)\,\rho_g}\right]^2 \cdot \left[\dfrac{(s_o - s_l)}{(s_g - s_l)} + \dfrac{(s_g - s_o)}{S^{*2}\,s_l}\right]} \right\}^{1/2} \tag{31}$$

Equation 31 is coupled, for the search for the maximum (critical) value of G, with Equation 30 and with the condition $\partial G/\partial p = 0$. The solution requires an iterative procedure.

The critical mass velocity may be obtained directly, depending on stagnation conditions, from diagrams of the type of Figure 5.[3]

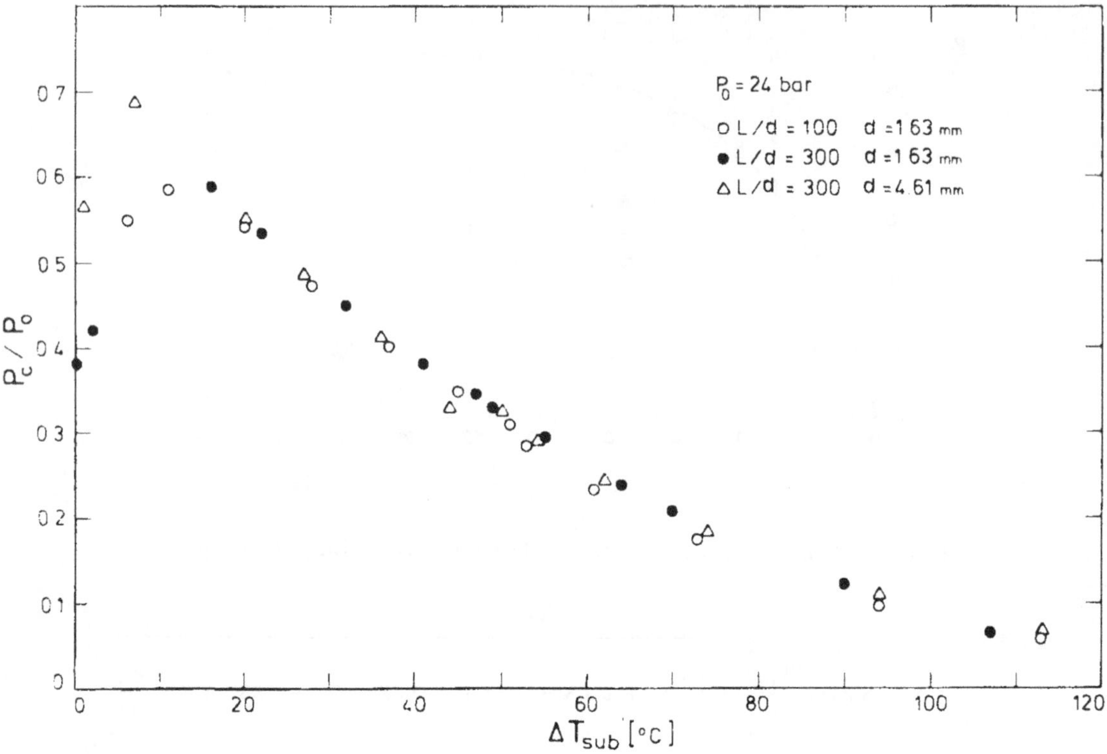

FIGURE 4. Critical ratio vs. inlet subcooling.

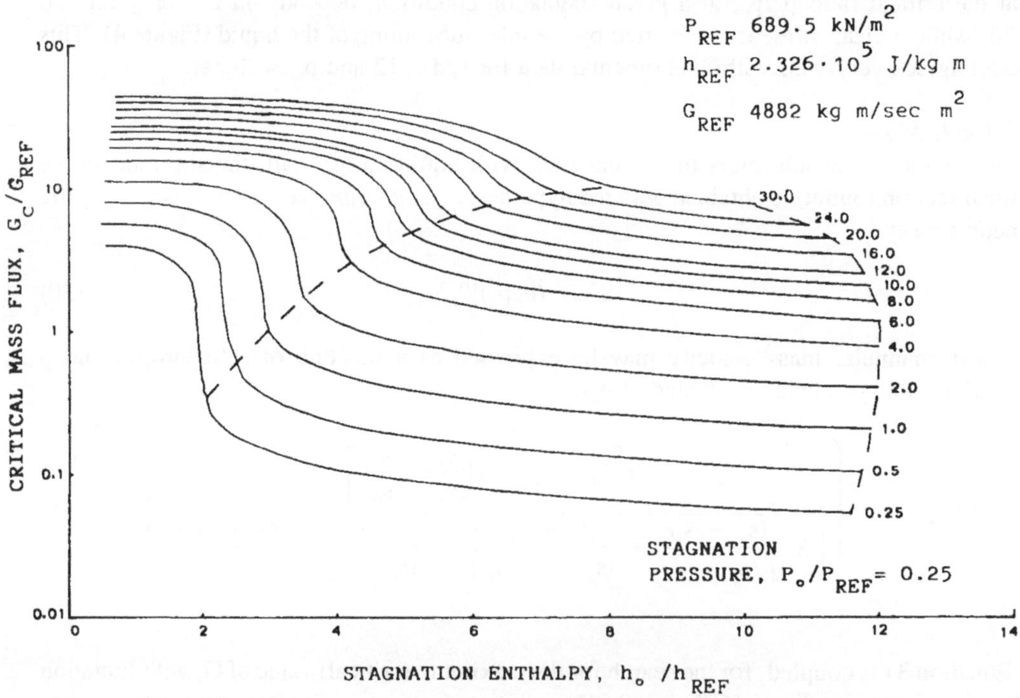

FIGURE 5. Critical mass velocity vs. inlet stagnation enthalpy and pressure (short channels).

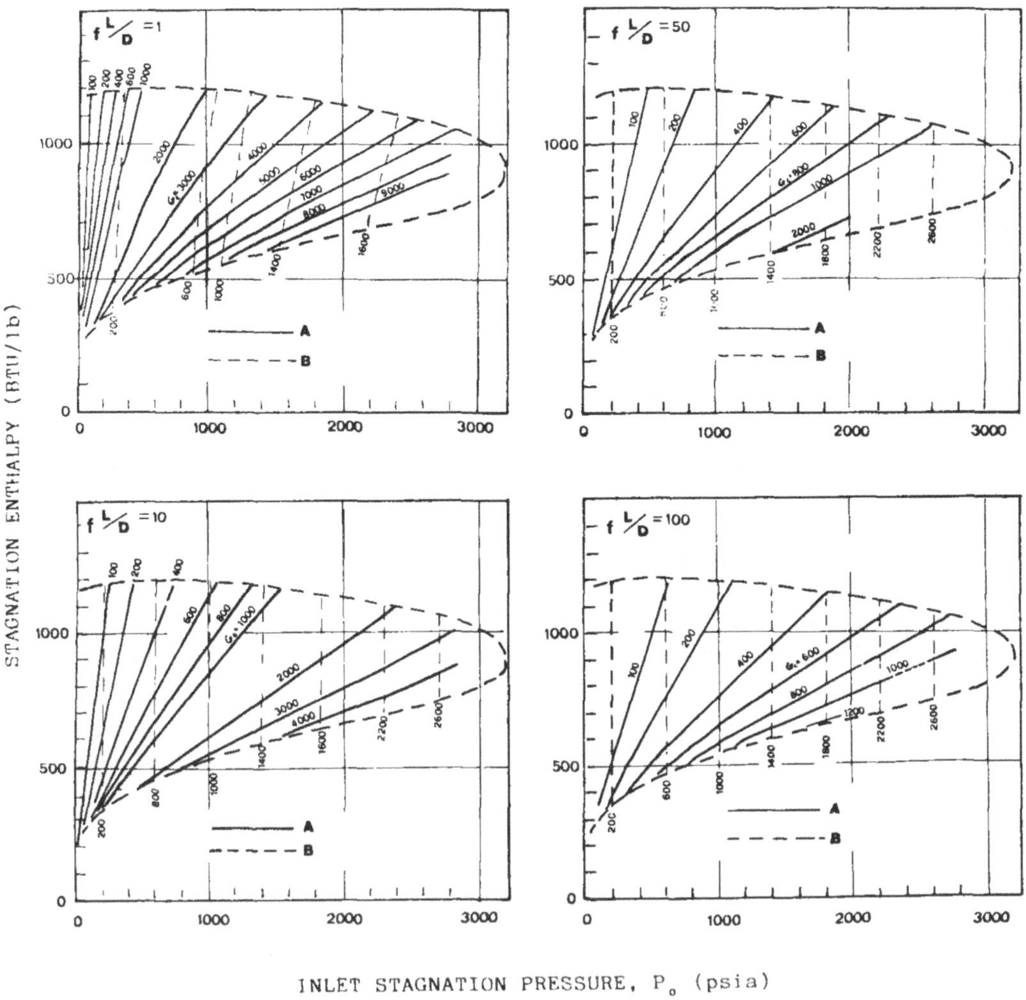

FIGURE 6. Critical mass velocity vs. inlet stagnation enthalpy and pressure. (A) Mass flow rate, lb/sec ft^2; (B) inlet static pressure, psia.

The Moody model is employed in nuclear safety analysis computer programs such as RELAP and RETRAN. With long channels, taking into account an isentropic expansion and introducing the friction effect between the liquid annular film and the solid wall, expressions of the type $G = G(p_o, i_o, f, L/d)$ are obtained for the critical mass velocity; the diagrams of Figure 6 plot the results for some typical situations.

B. Thermodynamic Disequilibrium Models

The models described in Section A assume that in sudden depressurizations the time intervals required for bubble formation and growth are negligible in comparison to transit times. In real flows, during a sudden depressurization, the liquid reaches a certain degree of superheating before the onset of nucleation; in critical-flow conditions, the time required may be comparable to the transit time. Metastability between the two phases has been apparent through experimental techniques; a typical situation is reported in Figure 7 showing the approximate pressure/time trend at the critical section.[4] Metastability times of the order of 10^{-4} sec have been revealed[4] in expanding jets through orifices in critical conditions. It is suggested[5] that the onset of vaporization is due to the simultaneous effects of a static expansion and of a turbulent fluctuation of pressure.

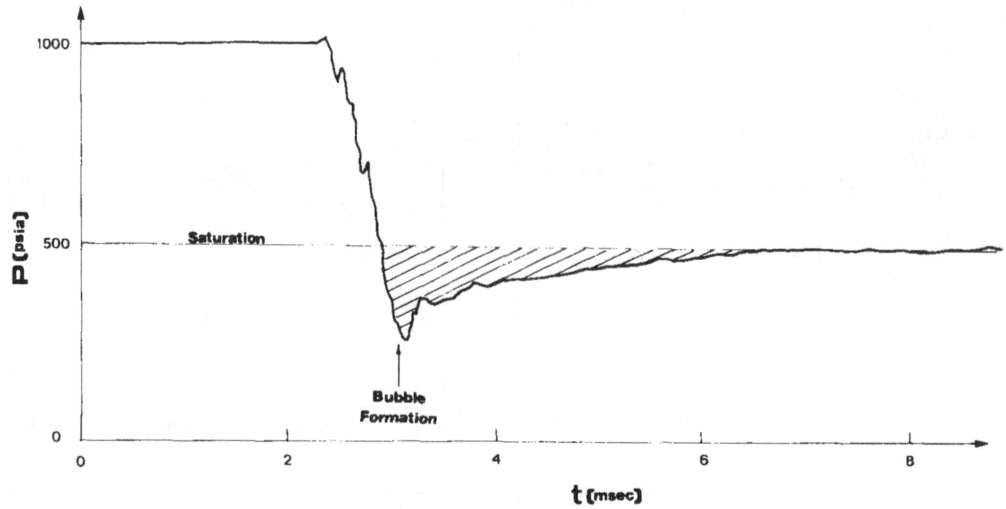

FIGURE 7. Near-outlet-section pressure trend.

1. Homogeneous Frozen Model[15]

In this model, the time required for vaporization is assumed to be greater than the mixture transit time in the channel. Other hypotheses are homogeneous flow; isentropicity of the expansion; and kinetic energy fully associated to vapor. Applying the energy equation from inlet (stagnation conditions) to the "critical" outlet section, the critical mass velocity is obtained:

$$G_{cr} = \rho\left(\frac{2\,X_o}{\rho_{go}}\,p_o\left(\frac{\gamma}{\gamma - 1}\right)\left[1 - \eta^{\left[\frac{\gamma-1}{\gamma}\right]}\right]\right)^{1/2} \tag{32}$$

where:

$$\eta = [2/(\gamma + 1)]^{\gamma/(\gamma - 1)} \quad \text{if} \quad [(1 - X_o)\,\rho_{go}]/[X_o\,\rho_{lo}] \ll 1$$

$$\text{and} \quad \rho = \left[(1 - X_o)\frac{1}{\rho_{lo}} + X_o\frac{1}{\rho_{go}}\,\eta^{\frac{1}{\gamma}}\right]^{-1}$$

2. Henry-Fauske Model[15]

This model simplifies the hypothesis of nonequilibrium assuming the onset of vaporization is always along the walls, as seen later in Reference 4 (see Figure 8).

The expression of the critical mass velocity is

$$G_{cr}^2 = -\left\{S\left[(1 + X(S - 1))\,X\,\frac{d\left(\frac{1}{\rho_g}\right)}{dp}\right. + \left[\frac{1}{\rho_g}(1 + 2X(S - 1)) + \frac{S}{\rho_l}(2(X - 1) + S(1 - 2X))\right]\frac{dX}{dp} + \right.$$
$$\left. + S\left[1 + X(S - 2) - X^2(S - 1)\right]\frac{d\left(\frac{1}{\rho_l}\right)}{dp} + X(1 - X)\left(\frac{S}{\rho_l} - \frac{1}{S\rho_g}\right)\frac{dS}{dp}\right]^{-1}\Bigg\} \tag{33}$$

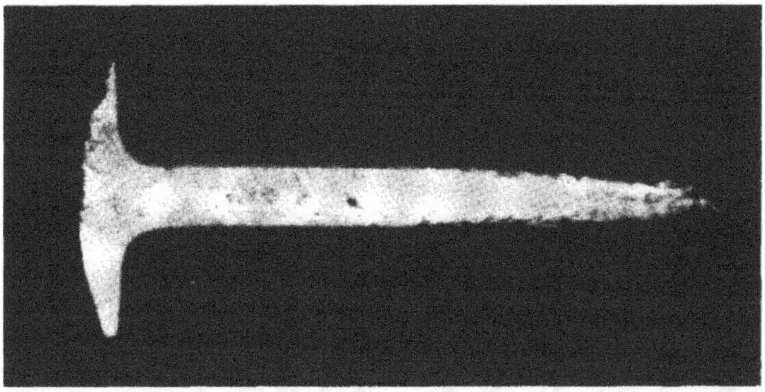

FIGURE 8. Typical example of the optical information obtained by means of Schlieren technique.

where the physical parameters are computed at the critical pressure. The following simplifications are proposed in the application of Equation 33:

1. Compressibility and specific volume of the vapor phase much greater than the liquid phase:

$$d(1/\rho_l)/dp \simeq 0; \quad 1/\rho_l \simeq 0$$

2. The local quality, X, expressed as a function of the corresponding thermodynamic-equilibrium quality X_e, according to the Henry model,[1] as follows:

$$N = \frac{X}{S\,X_e}; \quad \frac{dX}{dp} = SN\frac{dX_e}{dp} + SX_e\frac{dN}{dp} + NX_e\frac{dS}{dp}$$

$$S\,N\,X_e \ll 1$$

$$N = \begin{cases} X_e/0.14 & \text{if } X_e \leq 0.14 \\ 1 & \text{if } X_e > 0.14 \end{cases}$$

$$\frac{p_{cr}}{p_o} = \eta; \quad \eta = \left\{ \frac{\dfrac{1-\alpha_o}{\alpha_o}(1-\eta) + \dfrac{\gamma}{\gamma-1}}{\dfrac{1}{2\,\beta\,\alpha_{cr}^2} + \dfrac{\gamma}{\gamma-1}} \right\}^{\frac{\gamma}{\gamma-1}} \tag{34}$$

$$\beta = \left[\frac{1}{n} + \left(1 - \frac{\rho_{g,cr}}{\rho_{lo}}\right) \cdot \left(\frac{(1-X_o)\,N\,p_{cr}}{X_o(s_{g,e} - s_{l,e})_{cr}} \cdot \left(\frac{d\,s_{l,e}}{dp}\right)\right]_{cr} - c_{p,g}\left(\frac{\dfrac{1}{n} - \dfrac{1}{\gamma}}{s_{g,o} - s_{l,o}}\right)\right]$$

where α_o and α_{cr} are the void fraction in stagnant and critical conditions computed using the homogeneous model:

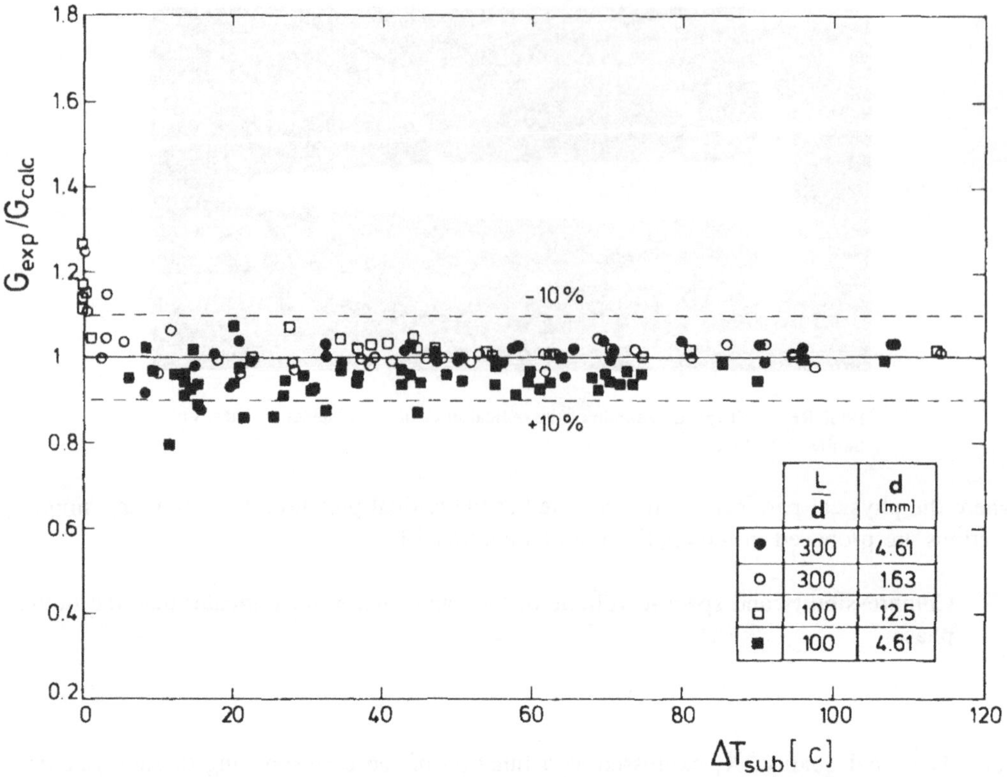

FIGURE 9. Comparison between Equation 12 and experimental data.

$$\alpha = (X/\rho_g)/[(1 - X)/\rho_l + X/\rho_g]$$

and the vapor critical density is computed as follows:

$$\left(\frac{1}{\rho_l}\right)_{cr} = \left(\frac{1}{\rho_g}\right)_o \left(\eta^{-\left(\frac{1}{\gamma}\right)}\right)$$

3. Modified Lackmé Model

Lackmé's original model[6] has been modified through "ad hoc" tests,[7] having good agreement with the experimental data (Figure 9). With reference to the flow channel of Figure 10, this model suggests the following expressions for the various pressure drop terms:

- Inlet pressure drop:

$$\Delta p_i = p_o - p_{sb} = K_{inlet}\, G_{cr}^2/(2\, \rho_o) \tag{35}$$

- Single-phase distributed pressure drop:

$$\Delta p_{sp} = p_{sb} - p_{nucl} = 4\, f(L_{sp}/d)\, G_{cr}^2/(2\, \rho_o) \tag{36}$$

- Two-phase distributed pressure drop:

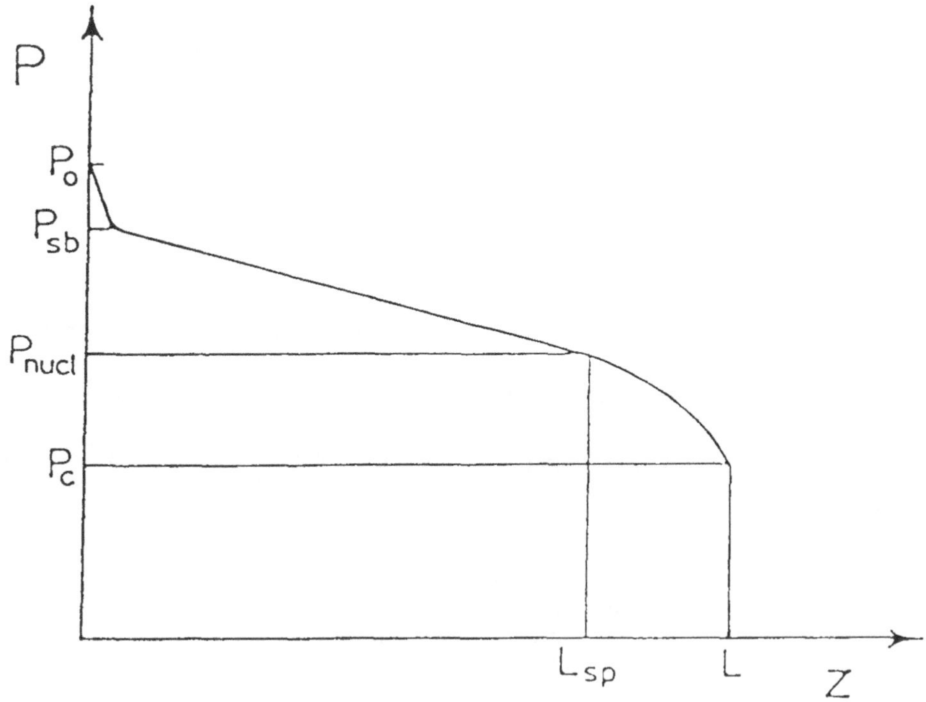

FIGURE 10. Pressure trend in a channel with critical flow.

$$\phi_{tp}^2 = 1 + 10(p_{cr}/p)\,X \tag{37}$$

$$\Delta p_{tp} = 4\,f(L_{tp}/d)\,\phi_{tp}^2\,G_{cr}^2/(2\,\rho_o) \tag{38}$$

where $L_{tp} = L - L_{sp}$. The two-phase multiplier $\phi^2{}_{tp}$ is computed at the following mean conditions: $X = X_{cr}/2$; $p = (p_{nucl} + p_{cr})/2$, p_{nucl} being a pressure suggested by Lackmé as $p_{nucl} = 0.92\ p_{sat}\ (T_o)$.

- Acceleration two-phase pressure drop:

$$\Delta p_{acc} = G_{cr}^2(1/\rho_2 - 1/\rho_1) \tag{39}$$

with:

$$\rho_1 = \rho_o; \quad \rho_2 = \left[\frac{X_{cr}^2}{\alpha_{cr}}\frac{1}{\rho_g} + \frac{(1 - X_{cr})^2}{(1 - \alpha_{cr})}\cdot\frac{1}{\rho_o}\right]^{-1} \tag{40}$$

The following relationships for the single-phase length of channel (L_{sp}) and the critical mass velocity (G_{cr}) are used:

$$L_{sp} = \frac{\left[\left(\frac{4f}{d}\right) L\,\phi_{tp}^2 + 2\left(\frac{\rho_o}{\rho_{cr}} - 1\right)\right](p_o - p_{nucl}) - K_i(p_{nucl} - p_{cr})}{\left(\frac{4f}{d}\right)[(p_{nucl} - p_{cr}) + \phi_{tp}^2(p_o - p_{nucl})]} \tag{41}$$

$$G_{cr}^2 = \frac{2\,\rho_o(p_o - p_{nucl})}{4f\,\frac{L_{sp}}{d} + K_i} \qquad (42)$$

The critical, outlet pressure is evaluated through the relationship[8]:

$$p_c = p_{cr} = p_{sat} - \frac{10^5}{20}\left(\frac{p_{sat}}{10^5}\right)^{[1.57\cdot(1 - p_{sat}/p^*)]} \qquad (43)$$

where p^* is the pressure at the thermodynamic-critical point and the pressure is expressed in Pa.

Quality in the outlet section of the channel (X_{cr}) is computed using the expression[1]:

$$X_{cr} = N\,X_e[1 - \exp(-B\,L_{tp}/d)] \qquad (44)$$

with:

$$N = \begin{cases} 20\,X_e & X_e < 0.05 \\ 1 & X_e \geq 0.05 \end{cases}$$

where X_e is the equilibrium quality, computed through the energy conservation equation and $B = 0.0523$ (empirical coefficient). For the void fraction computation, the Moody expression for the slip ratio, $S^* = (\rho_l/\rho_g)^{1/3}$, is adopted.

C. Separate-Phase Model

The major simplifications adopted in the models described in Section B consist in the assumptions concerning the momentum transfer between the phases and the delays due to metastability. The separate-phase models, adopting a much more detailed physical description (including mass, momentum, and energy transfer between the phases), have a much wider validity. They are, nevertheless, quite complex and for their practical application much more physical information is needed.

For the two phases, mass, momentum, and energy equations are written in a one-dimensional flow and the general assumptions are made that pressure in the two phases is equal in the same section and that the gas phase in the section is in saturation conditions. Usual unknowns are void fraction, quality, velocities, pressure, and the temperature of the phase in nonequilibrium conditions.

The differences among the proposed models are connected with the schematizations of thermal and mechanical interactions between the phases and with the definition of the critical condition. For a detailed analysis of some of the most relevant proposed models, see References 9 to 12. In particular, their main assumptions for the thermal interaction and for the critical conditions are considered here, while the mechanical interaction and the conservation equations are described in Chapter 4, Section III.C and Chapter 2, Section IV.E.3, respectively.

1. Thermal Interaction

According to Reference 10, the thermal power exchanged between the phases in the unit channel length (cross area A) of the phases may be expressed as:

$$d\dot{Q}/dz = \tilde{h}\,A_{lg}\,A(T_l - T_g) \qquad (45)$$

where A_{lg} is the two-phase interface per unit volume of the mixture and $\tilde{h}$ the interface heat transfer coefficient. Obviously, $\tilde{h}$ depends on the two-phase flow regime, with two definite models for bubbly flow ($\alpha \lesssim 0.3$) and annular flow ($\alpha \gtrsim 0.8$). For $0.3 < \alpha < 0.8$ an interpolation for $\tilde{h}$ values may be used (see also Chapter 2, Section IV.E.3 and Chapter 4, Section III.C):

- For a bubbly flow (situation I)[10]:

$$Nu = \tilde{h}\ d/zb/k_l = 2 + 0.6\ Re_b^{1/2}\ Pr_l^{1/3} \qquad (46)$$

with $Re_b = [\rho_l\ d_b\ (v_g - v_l)]/\mu_l$; Pr_l = liquid Prandtl number and d_b = bubble diameter.

- For an annular flow (situation II)[13]:

$$Nu = \tilde{h}\ d\ \alpha^{1/2}/k_g = 0.0023\ Re_g^{0.8}\ Pr_g^{0.4} \qquad (47)$$

where $Re_g = [\alpha^{1/2}\ \rho_g\ d\ (v_g - v_l)]/\mu_g$; Pr_g = vapor Prandtl number.

- For intermediate conditions:

$$\tilde{h} = \tilde{h}_I + (\tilde{h}_I - \tilde{h}_{II})\ (\alpha - \alpha_I)/(\alpha_I - \alpha_{II}) \qquad (48)$$

where $\tilde{h}I$ is determined from Equation 46 at $\alpha = \alpha_I = 0.3$ and $\tilde{h}II$ is found from Equation 47 at $\alpha = \alpha_{II} = 0.8$.

For the interface A_{lg}, the Richter expression of Chapter 4, Section III.C may be adopted.

Metastability conditions may be accounted for (Reference 10) assuming an active nuclei density of $1 \cdot 10^{11}$ nuclei per cubic meter and a nucleus diameter of $2.5 \cdot 10^{-5}$ m.

2. Critical Condition

The usual approach of the sonic velocity is of a difficult application in two-phase mixtures because the sonic velocities are very different in the two phases and the spatial distribution of phases greatly influences their value. In separate-phase models, the critical condition may be expressed writing the conservation equations in matrix form:

$$M_t\ \partial X/\partial t + M_z\ \partial X/\partial z = B \qquad (49)$$

(where M_t and M_z are the matrices of dependent variables derived with respect to time and space, X is the vector of the unknowns, and B is the vector of known terms) and then imposing the conditions: Det $(M_z) = 0$ and Det $(M_i) = 0$, M_i being the matrix obtained substituting to each column of M_z the column vector:

$$B - M_t\ \partial X/\partial t \qquad (50)$$

The condition Det $(M_z) = 0$ corresponds to the condition that the propagation velocity of pressure waves must be equal to zero, while the condition Det $(M_i) = 0$ corresponds to the assumption that in the critical section the spatial gradient of the X unknowns is undetermined. This exact approach may be simplified assuming as critical the section where $\partial p/\partial z > 1 \cdot 10^{10}$ Pa/m.[14]

REFERENCES

1. **Henry, R. E.,** The two phase critical discharge of initially saturated or subcooled liquid, *Nucl. Sci. Eng.*, 41, 1970.
2. **Fauske, H. K.,** The discharge of saturated water through tubes, *J. Heat Transfer,* 61(59).
3. **Lahey, R. T. and Moody, F. J.,** Thermal hydraulics of a boiling water nuclear reactor, American Nuclear Society, LaGrange Park, Ill., 1977.
4. **Celata, G. P., Cumo, M., Farello, G. E., and Incalcaterra, P. C.,** Critical Flow of Subcooled Liquid and Jet Forces, ENEA-RT/Inq., 82, 18, 1982.
5. **Jones, O. C. and Saha, P.,** Volumetric vapour generation in non equilibrium two phase flow, Advanced Code Review Group Meeting of Water Reactor Safety Research Division, U.S. Nuclear Regulatory Commission, Washington, D.C., 1977.
6. **Lackmé, C.,** Thermodynamics of critical two-phase discharge from long pipes of initially subcooled water, in *Heat Transfer in Nuclear Reactor Safety,* Bankoff, S. G. and Afgan, N. H., Eds. Hemisphere-McGraw Hill, New York, 1980.
7. **Celata, G. P., Cumo, M., D'Annibale, F., and Farello, G. E.,** Un metodo di calcolo per la valutazione della portata nell'efflusso critico di liquido inizialmente sottoraffreddato, paper presented at 40th Congr. Natl. ATI, Trieste, Italy, September 1985.
8. **Flinta, J.,** Critical flow in reactor safety analysis, paper presented at Eur. Two-Phase Flow Group Meet., Zurich, June 1983.
9. **Giot, M. and Fritte, A.,** *A Modeling of Critical Flow Advances in Two Phase Flow and Heat Transfer,* Vol. 1, Martinus Nijhoff, Boston, 1983.
10. **Richter, H. J.,** Separated Two-Phase Flow Model Application to Critical Two-Phase Flow, EPRI NP-1800 RP 443-2, 1981.
11. **Ardron, C. M.,** A two-fluid model for critical vapour-liquid flow, *Int. J. Multiphase Flow,* 4, 1977.
12. **Rivard, W. C. and Travis, J. R.,** A nonequilibrium vapour production model for critical flow, *Nucl. Sci. Eng.*, 74, 1980.
13. **Solbrig, C. W., McFadden, J. H., Lyczkowsky, R. W., and Hughes, B. D.,** Heat transfer and friction correlations required to describe steam-water behaviour in nuclear safety studies, *AIChE Symp. Ser.*, 174, 1978.
14. **Nguyen, D. L., Winter, E. R. F., and Ereiner, M.,** Sonic Velocity in Two-Phase System, University of Munich.
15. **Henry, R. E. and Fauske, H. K.,** The two-phase critical flow of one-component mixtures in nozzles, orifices, and short tubes, *J. Heat Transfer,* May 1971.
16. **Henry, R. E., Grolmes, M. A., and Fauske, H. K.,** Propagation velocity of pressure in gas-liquid mixtures, paper presented at Gas-Liquid Flow Symp. Waterloo Univ., Waterloo, Ontario, Canada, 1968.

Chapter 7

TWO-PHASE FLOW INSTABILITIES

I. INTRODUCTION

Some nuclear power plants and many heat exchangers where boiling or condensation occurs are subject to thermal-hydraulic instabilities. These may cause excursions or oscillations in the flow, which may induce thermal crisis, disturbances to the control systems, or damage of the mechanical type (thermal cycles). Various kinds of instabilities are examined in this chapter, according to a classification based on the related physical mechanisms. Some of the most effective analytical techniques employed to predict the instability thresholds are pointed out.

Thermodynamic instabilities are undesirable and potentially harmful in boiling and condensation processes and generally in the components with a two-phase flow.

The following consequences are possible:

1. Fluidodynamic oscillations of great amplitude may induce vibrations into the components
2. Flow and/or pressure oscillations may cause problems in the process control system
3. Flow oscillations may influence the local heat transfer characteristics, with consequent oscillations in the wall temperatures; the power limits at which the thermal crisis occurs may be substantially reduced, with damage to the heat exchange apparatus

The problems of flow stability are of particular importance in the nuclear reactors cooled and/or moderated with boiling light water and in steam generators (of the kinds used in the electric power generation plants, but, most of all, of those used in the chemical industry where the production of steam is associated to the fast cooling of a process gas).

Plant designers must predict the threshold for the onset of flow instabilities, in order to exclude them in the normal operation conditions or in order to reduce their effects by means of the feedback of appropriate control systems. It is not yet possible to schematize in a rational and exhaustive picture the various kinds of flow instabilities. In fact, some of them often coexist in the same operational situation; the criteria used for the classification are rather arbitrary and contradictory; above all, the phenomenon of the instabilities is not yet completely known in all its manifestations.

A relevant trial to rationalize and classify the instabilities has been performed in Reference 1, where some classification criteria, based on the physical mechanisms which seem responsible for the single instability, are proposed. Such criteria, adopted in this book, are still the most authoritative.

Analysis techniques are also proposed for some kinds of instabilities; since the behavior of a specific instability and the method for its prediction are different from those of other instabilities, the designers must initially be able to recognize the kind of flow instabilities that might occur, before applying the specific analysis.

II. FLOW INSTABILITY CLASSIFICATION

The proposed classification is based on the following definitions:

1. A *steady flow* is characterized by system parameters which are functions of the space variables only. In practice, the fluctuations usually occurring as an effect of turbulence, nucleation phenomena, or of the presence of a plug flow, do not modify the funda-

mentally stationary nature of the flow. They may, however, be able to initiate different instability phenomena.

2. A flow is *stable* when, momentarily disturbed, the new operation conditions tend asymptotically to the initial values.
3. A flow is subject to a *static instability* when, as a consequence of small variations in the stationary flow conditions, it is not possible to reach a new stationary situation in proximity of the preceding one. The cause of a static instability may be searched for in the trends or in the laws of the steady flow which govern the specific process (in particular, the instability threshold may be identified using the steady flow laws). A static instability may lead either to a different steady-state condition or to an oscillatory behavior, periodic in nature. In the latter case, the static instability behaves as the initiating event of a dynamic instability.
4. A flow is subject to a *dynamic instability* when the inertial terms and other feedback effects have a fundamental importance in the process. The fluid system parameters (pressure, flow rate, etc.) are subject to fast oscillations, so that it behaves as a servomechanism and the knowledge of the stationary flow laws is not sufficient even to determine the instability threshold; a dynamic instability may also be initiated only by the statistical fluctuations of the main parameters of the two-phase system.
5. An instability may be of *simple type* (or fundamental or pure), when a fundamental mechanism may be recognized, or of *complex* (or compound) *type*, when many elementary mechanisms of different natures interact contemporaneously, without any possibility of separating their effects.
6. An instability phenomenon is called *secondary* if it occurs as a consequence of another phenomenon, defined as *primary*. In this work, the item "secondary phenomenon" will be used only when the primary phenomenon is a necessary condition for the occurrence of the secondary one. A schematic classification of the different instabilities is indicated in Table 1.

III. PHYSICAL MECHANISMS AND RELEVANT EFFECTS

The following parameters must be considered *a priori* as important in order to control a process potentially unstable:

1. Geometry (length of the channel and/or of the loop, flow area, flow restrictions at the inlet or at the outlet, number of parallel channels)
2. Operation conditions (pressure, subcooling of inlet flow, mass flow rate, thermal power, natural or forced circulation)
3. Boundary conditions (axial heat flux distribution, flow or pressure imposed to the channel and/or to the circuit)

The effects of geometry and of boundary conditions are usually interconnected, as in the case of a flow distribution among many channels in parallel. Since parallel channels are connected at their ends to common headers, the flow distribution among the channels is determined by the dynamic answer of the single channel to the common boundary conditions, corresponding to the difference of pressure between the inlet and the outlet headers. The boundary conditions allow the separation of, in the analysis, the instability of the single channel from the eventual instability of the complex system. The physical mechanisms of the various kinds of flow instabilities are briefly described in the following paragraphs.

Table 1
TWO-PHASE FLOW INSTABILITY CLASSIFICATION

Category	Subcategory	Type	Mechanism	Characteristics
Static	Simple	Ledinegg (flow excursion) instability	$\left(\frac{\partial \Delta p}{\partial G}\right)_{int} < \left(\frac{\partial \Delta p}{\partial G}\right)_{ext}$	Flow undergoes a sudden, large amplitude excursion, to a new, stable operating condition
		Thermal (boiling) crisis	Substantial decrease of heat transfer coefficient	Wall excursion with possible flow oscillation
		Flow-regime transition instability	Bubbly flow has less void but higher Δp than annular flow; condensation rate depends on flow-regime	Cyclic flow-regime transitions and flow rate variations
		Nonequilibrium-state instability	Transformation wave propagates along the system	Recoverable work disturbances and heating disturbances waves
	Compound	Unstable vapor formation (bumping, geysering, vapor burst) (relaxation instabilities)	Periodic adjustment of a metastable condition, usually due to lack of nucleation sites	Occasional or periodic process of liquid superheat and violent vaporization with possible expulsion and refilling
		Condensation chugging	Bubble growth and condensation followed by surge of liquid (in steam discharge pipes)	Periodic interruption of vent steam flow due to condensation and surge of water up to downcomer
Dynamic	Simple	Acoustic oscillations	Resonance of pressure waves	High frequency pressure oscillations (10—100 Hz) related to time required for pressure wave propagation in system

Table 1 (continued)
TWO-PHASE FLOW INSTABILITY CLASSIFICATION

Category	Subcategory	Type	Mechanism	Characteristics
		Density-wave oscillation	Delay and feedback effects in relationships among flow rate, density, pressure drops	Low frequency oscillations (~1 Hz) related to transit time of a mass-continuity wave
	Compound	Thermal oscillations	Interaction of variable heat transfer coefficient with flow dynamics	Occurs close to film boiling
		Boiling water reactor instability	Interaction of void/reactivity coupling with flow dynamics and heat transfer	Relevant only for a small fuel time constant and under low pressure
		Parallel channel instability	Interaction among a small number of parallel channels	Various modes of dynamic flow redistribution
		Condensation oscillation	Interaction of direct contact condensation interface with pool convection	Occurs with steam injection into vapor suppression pools
	Compound as a secondary phenomenon	Pressure drop oscillation	A flow excurision initiates a dynamic interaction between a channel and a compressible volume	Very low frequency periodic process (~0.1 Hz)

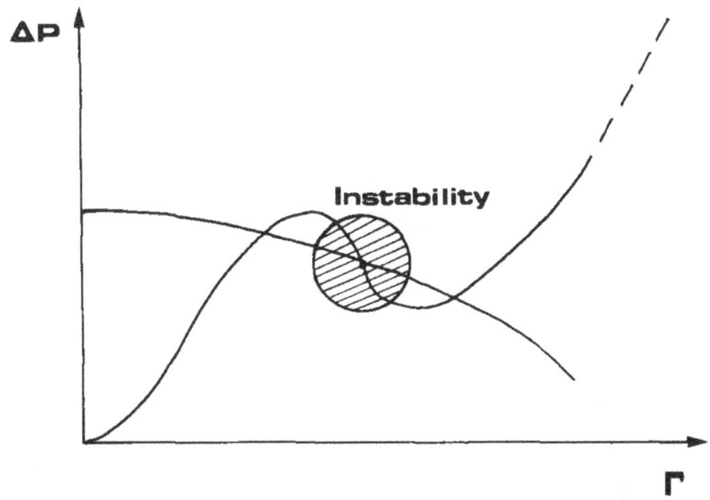

FIGURE 1. Characteristics of a circuit potentially subject to a static instability.

A. Static Instabilities

1. Simple Instabilities

a. Flow Excursion (Ledinegg) Instability

The flow excursion instability (or Ledinegg instability)[2] involves a sudden change of the flow to a considerably lower value. This happens when the slope of the characteristic curve — channel pressure losses vs. flow rate (channel internal characteristic) — becomes algebrically lower than the slope of the characteristic curve — motive-head vs. flow rate of the feeding component or system (channel external characteristic). The threshold criterion for this kind of instability is represented in Figure 1 and is expressed by the relationship:

$$(\partial\ \Delta p/\partial G)_{int} < (\partial\ \Delta p/\partial G)_{ext} \qquad (1)$$

This behavior requires that the channel internal characteristic has a region where the overall pressure losses (sum of friction, acceleration, and gravity terms) decrease with the increase of the flow. This situation may physically occur in a heated channel with high values of thermal fluxes in correspondence to the transition between nonboiling and subcooled boiling. In a heated channel with a forced convection liquid flow, if the flow rate decreases while the other parameters remain constant (in particular, heat flux) and the subcooled boiling conditions are reached, the pressure losses may grow considerably. This is an effect of the increase of friction and acceleration, due to the presence of voids in thermodynamic disequilibrium, especially in proximity of the heated wall. If the pump characteristic has a slope less negative then that of the channel, the system is unstable. With a pump characteristic of type A (Figure 2), for example, an operation in point b is unstable because a small decrease in the flow causes a spontaneous shift to point a. The new equilibrium point corresponds to a flow rate so low that the channel may be brought to thermal crisis conditions. With the introduction of an upstream valve in the heated channel, the global characteristic of the external pump (and of the valve) may become sufficiently sharp (B curve) so to avoid flow excursions and permit an operation in stable conditions in the region with negative slope (point d).

If a parallel-tube bundle is operated in conditions of subcooled boiling, Ledinegg instability may represent the limiting operation condition for the tubes, because each tube sees an essentially constant driving Δp (not negative slope) for the external characteristic. A stable

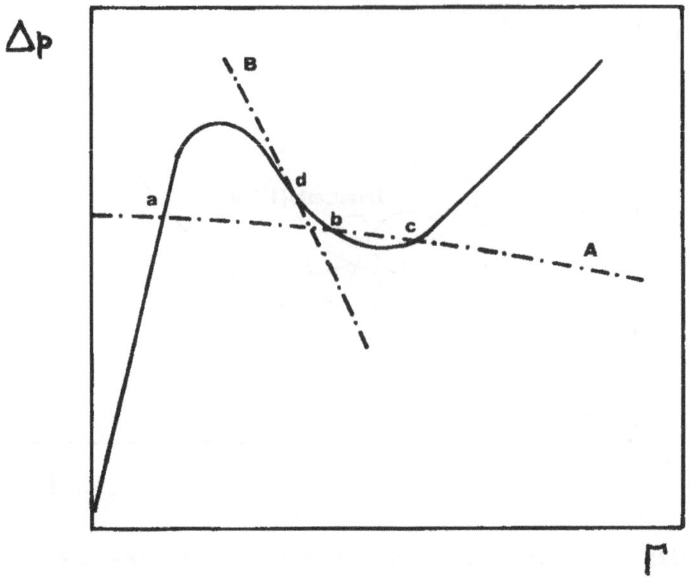

FIGURE 2. Characteristic Δp-Γ curves in a constant-flux heated channel.

operation may be obtained with the introduction of concentrated pressure drops on each parallel channel, in the inlet end, by means of throttling valves. Sometimes the increase required in the feeding pressure may be considerable.

In steam generators for liquid metal fast-breeder reactors, instability conditions have been identified in a preliminary theoretical analysis, but have not been verified experimentally; other parameters may influence the stability of such systems, where the thermal coupling between the primary and the secondary fluids has a stabilizing effect. In some situations, excursions have been verified with a very long excursion transient, due to the considerable length of the tubes.

b. Thermal Crisis

Thermal crisis is caused by a change in the heat transfer mechanism and is characterized by a sudden increase in the wall temperature. Many correlations have been proposed to interpret hydrodynamics and heat transfer mechanisms in proximity of a heating wall in thermal crisis conditions: some of these consider the thermal crisis as an instability phenomenon due to the separation of the liquid layer from the wall. This leads to a wall temperature increase, which may be associated to flow oscillations. For a deeper analysis of thermal crisis phenomena, see Chapter 10.

c. Flow-Regime Transition Instabilities

Flow-regime transition instabilities have been hypothesized with flow conditions close to the transition between the bubbly and the annular flow. A temporary increment of the bubble population in a bubbly flow or even in a plug flow (for example, due to a temporary reduction in the flow rate) may modify the kind of flow, making it annular with a reduction in pressure losses. The exuberance of the driving Δp will cause an increase in the flow rate, with a return to the bubbly or to the plug flow. A cyclic transition mechanism between the two flow regimes may occur. Oscillatory phenomena of this kind are generally characterized by a periodic behavior, with limited amplitudes and the existence of a threshold. Pressure oscillations are generally associated to flow oscillations of a relative importance. It must be noted that vibrations may occur as a consequence of the fluctuations in the flow rate and in the pressure.

d. Nonequilibrium State Instability

When nonequilibrium states occur, as in the interior of a moving shock wave, disturbance waves propagate along the system, belonging to two main types: (1) recoverable-work disturbances (as the compression of a gas) and (2) heating (dissipative-work) disturbances. Experience has shown that, in a given system, recoverable work disturbances cause state changes with time intervals roughly corresponding to the propagation time.[5] Whether propagation effects are important, they should be evaluated before a system is chosen. The procedure involves an estimate of the propagation time, t_p, which is compared with the appropriate problem time, t_∞. If propagation occurs in a very short time with respect to the problem time, propagation effects are probably of minor importance. "Bulk flows" ($t_p \ll t_\infty$) are characterized by negligible propagation effects, whereas "propagation flows" ($t_p \sim t_\infty$) include the propagation of disturbances, which may be of a relevant magnitude. Propagation effects probably are not important in analyses when it is $t_p \leq 0.2\ t_\infty$.

2. Compound Instabilities

a. Compound Relaxation Instabilities

Situations exist where, inside a heated system, irregular formations of vapor take place, which sometimes assume an explosive form and lead to fast flow rate variations and to coolant expulsion. Such phenomena are generally repetitive, even if not necessarily periodic.

A first phenomenon of this kind (bumping) is the migration of the boiling boundary inside a heated channel. It occurs, in particular, in alkaline metal boiling at low pressures, when the temperature on the wall is at the limit between boiling and natural convection. A movement of the boiling region along the channel may be noted, an effect that disappears with the increase of the heat flux and of the system pressure.

A second phenomenon of this kind (geysering) is typical of configurations where vertical liquid columns exist, with closed top ends, heated at the base. If pressure is low, if thermal fluxes are not high, and if boiling is initiated at the base, a sudden vapor generation may take place due to the reduction of the hydrostatic head. Generally, the fast vapor generation causes a kind of "explosion," with expulsion of the coolant from the lower end of the channel. If new cold liquid enters the channel, a natural convection condition is restored; if a boiling condition recurs at the base, the cycle starts again.

A third mechanism of unstable vapor formation is the vapor burst. This instability is characterized by the sudden appearance and the fast increase of a vapor-phase volume within a liquid where high values of overheating have been reached. This phenomenon occurs more frequently with alkaline liquid metals and with fluorocarbons, because both of these classes of fluids have near-zero contact angles on common walls (low surface tension). With liquids characterized by a good wettability, all the large cavities on the surface may be flooded. This leads to a reduction in the number of nucleation centers and to the need of a high value of overheating to obtain boiling (see Chapter 9). Only with very high heat fluxes and high overheating conditions of the wall, may a formation of vapor be obtained; it occurs with explosions causing even the expulsion of the coolant from the channel.

Many other kinds of vapor formation instabilities have been noted, always characterized by the existence of a phenomenon which is critical with respect to any governing parameter; if at the design stage, a sensitivity analysis shows the possibility of the existence of different behaviors of some governing parameters (heat transfer, flow regimes, etc.), an instability of this kind may be expected or pointed out.

b. Condensation Chugging

An important application of direct contact condensation in the field of nuclear safety involves the introduction of a high-enthalpy fluid (steam) in a pressure-suppression pool containing a low-enthalpy fluid (cold water). This limits the increase of pressure in the upstream volume. With low vapor flows and/or very low pool temperatures, the introduction of the vapor through the injection pipe (downcomer) may cause the formation of a bubble below the bottom of the downcomer. Subsequently, the bubble condenses and water surges up the downcomer in water-hammer fashion. The pressure oscillations due to the repetitive process of fluid expulsion and of water introduction into the downcomer has caused significant loads on some boiling water reactor (BWR) containment systems.

Condensation chugging has not been completely studied; it depends considerably on the geometries adopted and on the operation conditions of the fluids. In some cases, the use of many downcomers with small diameters instead of few downcomers with large diameters has solved the problem.

B. Dynamic Instabilities

1. Fundamental Dynamic Instabilities

The basic mechanism of fundamental dynamic instabilities involves the propagation of disturbances inside a two-phase flow, transported by two kinds of waves: pressure (acoustic) waves or density (void) waves. In every real system, both kinds of waves are present and interact with each other, but their velocities generally differ within one or two orders of magnitude, allowing a distinction between them.

a. Acoustic Instabilities

The acoustic oscillations revealed in the majority of two-phase systems, due to the propagation of pressure waves in the coolant flow, are characterized by a high frequency (10 to 100 Hz), correponding to the period required by a pressure wave to travel through the system. Acoustic oscillations have been observed in subcooled, bulk, and film boiling. The pressure losses amplitude due to such oscillations may be relevant and the pressure fluctuations which occur at the inlet of the channels may represent a significant fraction of the average value of the pressure itself. The oscillation amplitude reduces, passing from subcooled boiling to bulk boiling. Other kinds of acoustic oscillations, in the range of 1000 to 10,000 Hz, may be observed in hypercritical flow conditions or in transition from subcritical to hypercritical pressure (cryogenic fluids).

Acoustic oscillations are potentially harmful or dangerous, leading to mechanical vibrations of the components. Their frequencies are so high, however, that in general, they do not pose any danger of thermal crisis or to the control system; no overlap exists with the frequency ranges of such phenomena. For the acoustic instabilities, empirical correlations have been proposed[7] to evaluate their thresholds and fundamental frequencies; such correlations are generally applicable only in the directly analyzed conditions and generalized criteria are not yet available.

b. Density-Wave Instabilities

These are the most important kinds of dynamic instabilities from the point of view of applications. They are characterized by mass flow rate variations in time and in space, through a system with a two-phase flow. As a consequence, the fluid density varies in function of time and space.

If a boiling channel is considered, a perturbation in the inlet flow rate modifies (with a constant thermal power) the density of the two-phase mixture. Variations in pressure losses follow, with feedback on the initial cause of perturbation (the flow rate).

Considering the system from a stationary point of view and excluding static-instability situations, the feedback has a stabilizing effect. It intervenes, in fact, on the cause with an opposite "direction" with respect to the cause itself (for example, an increase in the flow rate causes an increase in the pressure drops and this, in turn, reduces the flow rate).

From a dynamic point of view, on the contrary, it is important to consider that the flow perturbation (the density wave) does not spread instantaneously; the feedback intervenes on the cause with a delay. The time phase displacement between the cause and the effect may become such that the feedback intervenes on the perturbation when it has already changed its sign. The feedback itself assumes a destabilizing characteristic, with the exaltation of the perturbating cause. The result is oscillatory behavior of the flow with a smoothed, constant, or divergent trend, in the function of the specific configuration. The oscillation frequency may be evaluated in function of the propagation time of the density wave in the system.

In order to determine analysis criteria, the following aspects must be considered: a dynamic instability occurs when the phase displacement between the cause and the effect reaches 180°; depending on the ''gain'' of the system (amplitude of the effect in the function of that of the cause), the answer is divergent, constant, or smoothed.

The occurrence of a dynamic instability depends upon: (1) the presence of perturbations with suitable frequencies (which affect the phase-displacement) and (2) the characteristics of the system (its geometry, the presence of resonance frequencies, etc.). For what concerns (1), whatever fluid system (especially, a two-phase system) contains perturbation sources with a wide frequency spectrum (due to the feeding system, operation conditions variations, and boiling phenomena, etc.), therefore, being potentially unstable. For what concerns (2), most two-phase systems include a first section crossed by liquid and a second section containing the two-phase mixture.

The effect of the two sections in relation to the dynamic answer of the system is completely different; the single-phase fluid is a system with a very high rigidity, where the flow perturbation spreads ''instantaneously'' with very little time phase displacement, while the opposite case occurs in two-phase mixture.

Obviously, the answer to the single-phase section is always stable, while for the two-phase section, it is possible to have a destabilizing behavior. The overall system answer results superposing the two answers; it will be closer to instability conditions if the pressure drops in the two-phase section prevail on those of the liquid section. Consequently, the importance of the pressure drops in the single-phase section, before the boiling channel, may be understood.

It is possible, at last, to formulate some relevant considerations:

1. Natural circulation is a situation particularly susceptible to initiate a dynamic instability. Almost constant pressure drops between the inlet and the outlet of the channel occur; the only stabilizing effect is that of the pressure drop due to the eventual single-phase flow portion of the system.
2. Another situation susceptible to dynamic instability is that of parallel boiling channels. In fact, a variation of the flow rate in a single channel does not substantially modify the pressure difference between the inlet and the outlet, so that stabilizing feedbacks do not occur. Parallel boiling channels represent a particularly frequent situation in the industrial applications of two-phase mixtures (boiling water nuclear reactors, steam generators, thermal electric plant boilers, etc.).
3. The introduction of suitable pressure drops (orifices) in the single-phase section upstream for the heated channels is a very effective method for eliminating dynamic instabilities.
4. Both in forced and in natural circulation, a reduction of the heated channel length has a stabilizing effect.
5. An increase of the system pressure, with constant thermal power, reduces the void fraction and the pressure drops of the two-phase flow due to friction and to acceleration. These effects are similar to those of a reduction in the thermal power or of an increase in the flow rate, and stabilize the system.

6. An increase of subcooling at the channel inlet reduces the void fraction and increases the nonboiling length and the transit time. If subcooling is average or high, its increase has a stabilizing effect. With a low subcooling, on the contrary, the effect is of destabilization.
7. The increase of the mass flow rate and the reduction of thermal power, *ceteris paribus*, have a stabilizing effect.

2. Compound Dynamic Instabilities: Primary Phenomena

Under this definition a wide series of complex phenomena is included.

a. Thermal Oscillations

Temperature oscillations have been observed in heated walls in heat transfer conditions of postdry-out (Chapter 11). The phenomenon has been interpreted as a consequence of an oscillating behavior of the fluid system between a film boiling and a transition boiling situation. Such heat transfer mechanisms lead to wide oscillations of temperature in the heating wall. The phenomenon is sometimes complicated by an axial fluctuation of the thermal crisis occurrence point. Dry-out oscillations may lead to an effect of fatigue and to a corrosion damage on the wall.

b. BWR Instabilities

In nuclear BWRs, a particular mechanism of thermal-hydraulic instability occurs, due to feedback among the void fraction, the nuclear reactivity, and the generated power. A relevant destabilizing effect may occur when the time constant of a hydraulic oscillation (density oscillation) is close to the thermal time constant of fuel rods.

c. Parallel Channel Instability

This situation may occur in systems with heated parallel channels. The static criterium of the distribution of flow rates among the channels may lead to an evaluation not corresponding to the real situation, which also requires an analysis on the basis of dynamic criteria. In fact, owing to interactions among the inlet flows, the vapor generation rates and the counterpressions due to pressure drops, dynamic phenomena may be initiated. They lead to oscillations of the flow rates in the channels, with characteristic amplitudes, frequencies, and phases. At different time values, the system may be in conditions of different flow distributions and far from those computed using stationary criteria. Also, this kind of complex instability may be reduced or eliminated with the use of orifices at the inlet of the single channels. The instability condition is reached by both increasing the thermal power (at constant global flow rate) and decreasing the flow rate (at constant power). Figure 3 shows typical experimental results at the onset of instabilities, while Figure 4 shows the stability maps obtained employing two descriptive dimensionless groups: the subcooling number $N_{sub} = \Delta h_{sub}\,\Delta\rho/(\rho_g\lambda)$ and the phase-change number $N_{pch} = W\,\Delta\rho/(\lambda\Gamma\rho_g)$, where W is the thermal power, Γ the total mass flow rate, and λ the latent heat of vaporization.

These maps are obtained for particular values of the inlet throttling (Figure 5 shows the experimental apparatus, consisting of two parallel heated channels, with different heat flux axial profiles, crossed by two-phase mixtures), expressed by the characteristic coefficients, K_i, of the inlet valves: $K_i = 2\Delta p/\rho v^2$.

3. Compound Dynamic Instabilities: Secondary Phenomena

Phenomena of instability may occur generated by other instabilities (primary and secondary instabilities). An example of this kind may be found in the oscillation which occurs in heated channels connected to a compressible upstream volume. Also in this case, the insertion of a throttling valve between the compressible volume and the heated channel may stabilize the flow characteristics.

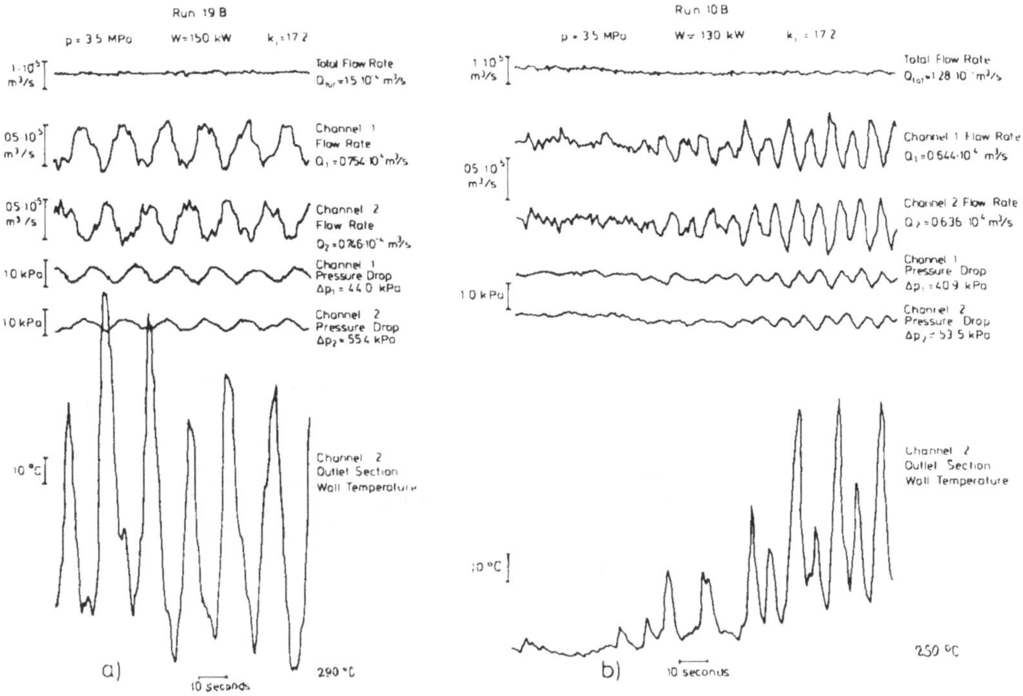

FIGURE 3. Typical experimental results showing instability onset in equipment of Figure 5.

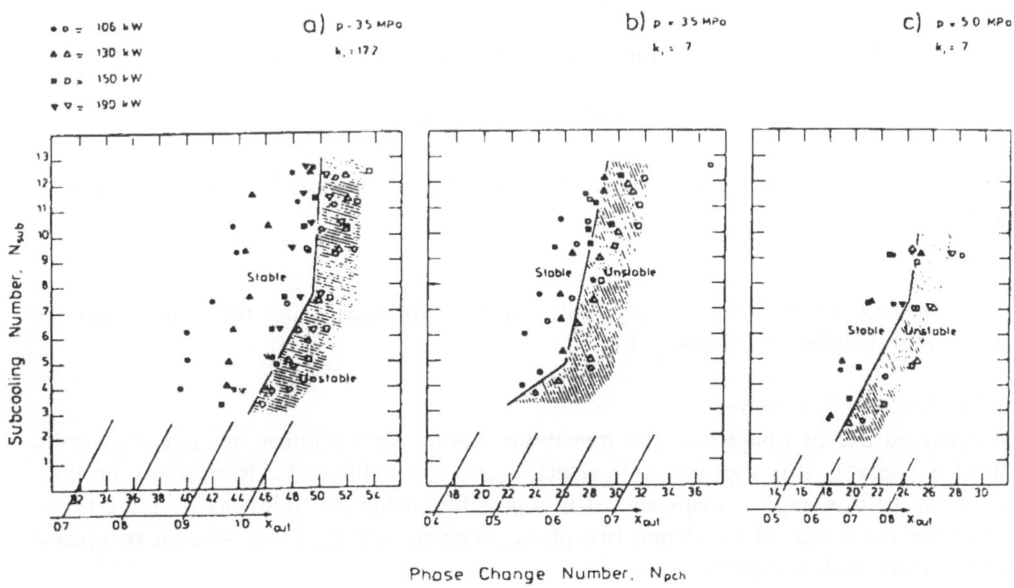

FIGURE 4. Typical experimental stability map.

IV. INSTABILITY ANALYSIS METHODS

A. Static Instabilities

The threshold of such instabilities may generally be foreseen using evaluations of a stationary kind.

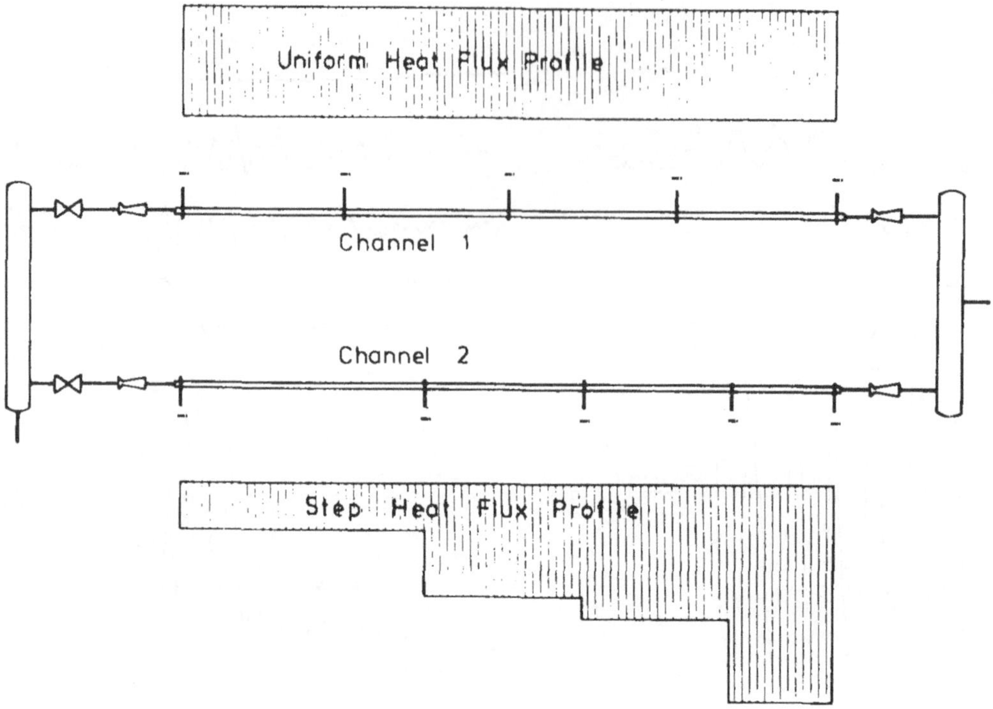

FIGURE 5. Parallel-channel-instability test section.

1. Flow Excursion

The threshold of flow excursion may be foreseen using the Ledinegg criterion:

$$(\partial\ \Delta p/\partial G)_{int} < (\partial\ \Delta p/\partial G)_{ext}$$

In the situation of several parallel channels, the verification must be done in the hottest channels.

2. Thermal Crisis

The threshold of this kind of instability must be foreseen using the criteria and the relationships analyzed in Chapter 10.

3. Flow-Regimes Transitions

As pointed out in Chapter 3, the transitions among flow regimes are not completely defined. Generally, it is appropriate to select operation conditions far from transition lines defined on the flow-regime maps and, if it is possible, using the following design criteria: (1) limiting the length of horizontal two-phase channels and (2) using vertical two-phase channels with small diameters.

B. Dynamic Instabilities

The analysis of two-phase flow dynamic instabilities is performed using three main methodologies: (1) empirical observation and phenomenological considerations, (2) direct integration of fundamental conservation equations using numerical methods, and (3) analytical solution of conservation equations.

The first method is based on the search for different effects, concomitant with the specific dynamic instability. A typical example is that of the density-wave oscillations, for which many phenomenological models have been developed. These models, have, most of all, a value in the introduction to the comprehension of the phenomenon.

The second method has a greater practical value. The direct numerical integration of conservation equations (not linear in the time domain) is often rather complex. Furthermore, attention must be given to the solution method because an inappropriate numerical method may show false oscillations, while the real solution is stable. Also, if only numerical instabilities (due to numerical solution methods) may sometimes be individuated, doubt may remain as to the real nature of the revealed instabilities (only numerical or real). Many computer programs have been developed to study the transient behavior, including instabilities, in two-phase flows.[1]

For the third method, there are many examples, even if there is no general solution for nonlinear differential equations. Linearized methods are, in fact, generally employed. They may predict the threshold for the instability and the decay modes of infinitesimal disturbances when the reference situation is stable, but may hardly predict the consequences of nonlinear effects.

At present, only numerical techniques are employed to study the effect of disturbances of great amplitude (nonlinear).

C. Analysis Phases

In order to perform an analysis, it is very important to understand and identify the physical fundamental mechanisms which may lead to instabilities in the specific system. As an example, it may be useful to consider the mechanisms of the acoustic oscillations and of density-wave oscillations. Many models for the density-wave oscillation's analysis neglect the effects of compressibility; the use of such models may lead to erroneous results if, in the specific system relevant acoustic waves, with interactions with density waves, occur.

At a design stage of a two-phase system the following scheme of sequential approach for the analysis of stability may be proposed:

1. Verification of the system with respect to Ledinegg instability. If the system is not stable, the following procedure may be adopted: to throttle the inlet; to increase the slope of the pump characteristic curve; to increase the downcomer hydraulic resistance in the situation of a natural recirculation loop.
2. Verification, through correlations, of the other potential static instabilities: thermal crisis, vapor burst, transitions among flow regimes.
3. Verification of the threshold of instability for density waves, using a simplified model with adimensional parameters, if the geometry and the boundary conditions of the system correspond to those used for the development of the adimensional models available.
4. Final verification of the instability threshold for density waves, using the most suitable computer program among those available for the considered situation.

REFERENCES

1. **Bouré, J. A., Bergles, A. E., and Tong, L. S.** Review of two-phase flow instability, *Nucl. Eng. Design*, 25, 1973.
2. **Ledinegg, M.,** Instability of flow during natural and forced circulation, *Waerme*, 61(8) 891, 1954.
3. **Cumo, M., Palazzi, G., and Rinaldi, I. L.,** An Experimental Study on Two-phase Flow Instability in Parallel Channels with Different Heat Flux Profiles, CNEN RT/ING, 81, 1, 1981.
4. **Bergles, A. E.,** Instabilities in two-phase systems, in *Two-Phase Flows and Heat Transfer*, Vol. 1, Hemisphere Publishing, Washington, D.C., 1977.
5. **Moody, F. J.,** personal communication, 1985.

many phenomenological models have been developed. These models have, however, all a value in the contribution to the comprehension of the phenomenon.

The second method has a greater practical value. The direct numerical integration of conservation equations (not linear in the time domain) is often rather complex. Furthermore, attention must be given to the solution method because an inappropriate numerical method may show false oscillations, while the real solution is stable. Also, if only numerical instabilities (due to numerical solution method) may sometimes be individualized, doubt may remain as to the real nature of such so-called instabilities (only numerical or real). Many computer programs have been developed to study the transient behavior, including instabilities, in two-phase flows.

For the second method, there are many examples, even if there is no general solution for nonlinear differential equations. Linearized methods are, in fact, generally employed. They may predict the threshold for the instability and the decay modes of an infinitesimal disturbance when the reference situation is stable, but may hardly predict the consequences of nonlinear effects.

At present, only numerical techniques are employed to study the effect of disturbances of great amplitude (nonlinear).

C. Analysis Phases

In order to perform an analysis, it is very important to understand and identify the physical fundamental mechanisms which may lead to instabilities in the specific system. As an example, it may be useful to consider the mechanisms of the acoustic oscillations and of density-wave oscillations. Many models for the density-wave oscillation analysis neglect the effects of compressibility: the use of such models may lead to erroneous results if, in the specific system, relevant acoustic waves, with interactions with density waves, occur.

At a design stage, of a two-phase system the following scheme of sequential approach for the analysis of stability may be proposed:

1. Verification of the system with respect to Ledinegg instability. If the system is not stable, the following procedure may be adopted: to throttle the inlets to increase the slope of the pump characteristic curve; to increase the downcomer hydraulic resistance in the situation of a natural recirculation loop.
2. Verification, through correlations, of the other potential static instabilities: thermal crisis, vapor burst, transitions among flow regimes.
3. Verification of the threshold of instability for density waves, using a simplified model with nondimensional parameters, if the geometry and the boundary conditions of the system correspond to those used for the development of the nondimensional models available.
4. Final verification of the instability threshold for density waves, using the most suitable computer program among those available for the considered situation.

REFERENCES

1. Boure, J. A., Bergles, A. E., and Tong, L. S., Review of two-phase flow instability, Nucl. Eng. Design, [illegible], 1973.
2. Ledinegg, M., Instability of flow during natural and forced circulation, Warme, 61(8), 891, 1938.
3. Cumo, M., Palazzi, G., and Urbani, G. C., An Experimental Study on Two-phase Flow Instability in Parallel Channels with Different Lengths, Rep. CNEN-RT/ING(81) 1, 1981.
4. Bergles, A. E., Instabilities in two-phase systems, in Two-Phase Flows and Heat Transfer, Vol. 1, Hemisphere Publishing, Washington, D.C., 1977.
5. Mosely, [illegible], personal communication, 1982.

Chapter 8

SINGLE-PHASE FORCED CONVECTION HEAT TRANSFER

I. INTRODUCTION

The single-phase-fluid heat transfer generally consists of two contributions: conduction through the fluid and heat transport associated to the motion of fluid particles between the wall layer and the bulkstream. The importance of the two contributions depends on the fluid thermal conductivity and on the flow turbulence level. A high level of turbulence leads to a high degree of mixing among molecules with different kinetic energy (thermal agitation), determining a high capacity of heat transport from/to the solid surface.

In order to determine the heat transfer coefficient, conservation equations should be solved which are necessary for the evaluation of the velocity and temperature profiles close to the solid surface, on which the coefficient depends (see Chapters 2 and 5).

The solution of conservation equations is obtained analytically only in the particular situation of laminar flows, with quite simple geometries (flow parallel to slabs, flow inside tubes with a completely developed velocity profile).

In the revelant situation of a laminar flow in a horizontal tube (pipe), if the velocity profile is completely developed* and radial density variations may be neglected, the velocity profile is given by (τ_w is the shear stress on the wall):

$$v(r) = \frac{\tau_w}{\mu\, d}\left[\left(\frac{d}{2}\right)^2 - r^2\right] = 2\,\bar{v}\left[1 - \frac{r^2}{d^2/4}\right] \qquad (1)$$

while the volumetric flow rate Q is given by:

* After a flow enters a duct with constant transversal geometry (i.e., constant diameter and straight axis for a tube), its velocity profile stabilizes within a length, called asset length. In laminar flows inside tubular ducts,the asset length (l_a) has been evaluated through the relationship:

$$\frac{l_a}{d} = 0.0575 \cdot Re$$

In turbulent flow, a theoretical study[1] has led to the relationship:

$$\frac{l_a}{d} = 4.63\, Re^{0.12}$$

A subsequent experimental study,[2] referring to ducts with an inlet section both with sharp and rounded edges, has led to the following relationships:

- Sharp-edge inlet:

$$\frac{l_a}{d} = 0.094\, Re^{0.48} \qquad Re \leq 1 \cdot 10^5$$

- Rounded-edge inlet:

$$\frac{l_a}{d} = 0.008\, Re^{0.685} \qquad Re \leq 1 \cdot 10^5$$

$$Q = \pi\left(\frac{d}{2}\right)^2 \bar{v} = \frac{\pi\ \tau_w\ d^3}{32\ \mu} \tag{2}$$

If the flow is turbulent, the solution of conservation equations is still possible only with quite simple geometries and with some basic simplifying assumptions (see Chapter 2). Such approximations allow particularly the definition of the velocity field near the wall, from which the temperature profile may be deduced.

With specific reference to the relevant applicative situation of a single phase turbulent flow inside a tube with a completely developed velocity profile, using the Prandtl assumptions for the expression of Reynolds stresses, conservation equations may be solved leading to the following expression for the velocity profile (universal velocity profile):

$$v^+ = y^+, \quad \text{if} \quad y^+ \leq 5 \text{ (in proximity to the wall)} \tag{3a}$$

$$v^+ = a_1 \ln y^+ + a_2, \quad \text{if} \quad y^+ > 5 \tag{3b}$$

with

$$v^+ = \frac{\bar{v}}{\sqrt{\dfrac{\tau_w}{\rho}}} \quad \text{and} \quad y^+ = y\,\frac{\sqrt{\tau_w/\rho}}{\nu}$$

y being the distance from the wall ($y = d/2 - r$). It is $a_1 = -3.05$, $a_2 = 5$ if $5 < y^+ \leq 30$ and $a_1 = 5.5$, $a_2 = 2.5$ if $y^+ > 30$. On the basis of the velocity profile, if the energy equation is solved, the universal temperature profile may be obtained*:

$$T^+ = \int_0^{y^+} \frac{d\,y^+}{\dfrac{1}{Pr} + \dfrac{\nu_t}{(\nu\,Pr_t)}} \tag{4}$$

where

$$\frac{\nu_t}{\nu} = (0.4\,y^+)^2 \cdot [1 - \exp(-y^+/26)] \cdot \left|\frac{\partial v^+}{\partial y^+}\right|;$$

$$T^+ \equiv \frac{(T_w - T)\,\rho\,c_p\,\sqrt{\dfrac{T_w}{\rho}}}{q''_w}$$

In common applications, the geometry often is not tubular and a completely developed velocity profile may not always be assumed.

This leads to the necessity of the empirical determination of heat transfer coefficients.

In such situations, conservation equations are still used, not for an analytical solution of the thermal problem, but for the possible identification of the most relevant parameters or groups of parameters to be used in the formulation of the empirical relationships (see Chapter

* As pointed out in Chapter 5, completely developed turbulent flows are characterized by a laminar boundary layer near the wall, where heat is transferred by conduction. In heat transfer correlations, great care is devoted to the laminar layer (laminar film) and to its temperature T_f (it is often assumed $T_f = (T_w + T_b)/2$). Physical properties are usually evaluated at T_f, even if practical reasons make the properties evaluation at T_b (temperature representative of the greater part of the fluid and relevant in energy balances) preferable.

5). A very effective method used in these cases for the determination of heat transfer coefficient is the dimensional analysis. It allows the identification of a functional dependence among the most relevant geometrical and physical parameters; it requires, however, the determination of empirical factors and this is possible only through specific experimental analyses.

In this chapter, the relationships applicable to the most common geometrical situations are reported, independent of the specific method used to obtain them.

II. FLOW INSIDE TUBES

Several relationships are available, referring to a great number of experimental tests performed because of the industrial relevance of this geometrical situation. The various relationships proposed for heat transfer coefficients show substantially the same fundamental structure; their most relevant differences involve empirical coefficients which are peculiar of the particular geometries or of the ranges of the most important physical fluid parameters.

With turbulent flows and $Pr \simeq 1$:

$$Nu = a_1 \cdot Re^{a_2} \cdot Pr^{a_3} \tag{5}$$

If the flow is laminar (Re <2500) (situation of applicative interest mainly with high-viscosity fluids or with liquid metals):

$$Nu = a_1 + a_2 \cdot (Re \cdot Pr)^{a_3} \tag{6}$$

where a_1 takes into account the great importance of the conductive effect in such a region. This parameter is strongly dependent on the nature of the liquid and, in particular, on its conductivity and degree of purity (mainly for liquid metals). The coefficient a_2 and the exponent a_3 are, on the contrary, strongly dependent on the wettability of the solid wall, which depends on its finishing conditions, the fluid physical characteristics, and the thermal gradient to which the fluid is subjected in proximity of the wall.

Equations 5 and 6, valid for a completely developed flow, must be modified to take into account the presence of entrance effects, bends, or other situations which prevent a complete development of the velocity profile. In the following, the most relevant applicative situations are considered.

A. Laminar Flow

In the region of a completely developed velocity profile (parabolic; inside a tube) in the hypotheses of constant physical properties, heat transfer by conduction, and T_w = const, Graetz proposed[3]:

$$\frac{T_2 - T_1}{T_w - T_1} = 1 - 8\, f(Gz^*) \tag{7}$$

where T_1 and T_2 are the inlet and outlet temperatures. If L is the tube heated length:

$$Gz^* = \frac{\pi}{4} \frac{k}{\mu\, c_p} \cdot \frac{\mu\, L}{\Gamma};$$

$$f(Gz^*) = 0.10238\, e^{-14.627 Gz^*} + 0.0122\, e^{-89.22 Gz^*} + \ldots$$

If an average heat transfer coefficient is introduced, using the following:

$$\pi d\, L\, \bar{h}\, \frac{[(T_w - T_1) + (T_w - T_2)]}{2} = \Gamma\, c_p(T_2 - T_1)$$

the Graetz correlation may be modified as follows:

$$\frac{\bar{h}\, d}{k} = \frac{2}{\pi} \cdot Gz \cdot \frac{1 - 8\, f(Gz^*)}{1 + 8\, f(Gz^*)} \tag{8}$$

which may be approximated by the empirical relationship:

$$\frac{\bar{h}\, d}{k} = 1.75\, (Gz)^{1/3} \quad \text{if} \quad Gz > 10 \tag{9}$$

With low values of $Gz(<20)$, it is simply:

$$Nu = 3.66 \tag{10}$$

In the case of a uniform heat flux:

$$Nu = 4.36 \tag{11}$$

In the entrance region, where the velocity profile is not yet completely stabilized (asset region), the average heat transfer coefficient (between 0 and z) is evaluated making use of the relationship:

$$Nu = 1.62 \left(\frac{\pi}{4}\, Re \cdot Pr\, \frac{d}{z}\right)^{1/3} \tag{12}$$

valid for T_w = const, with $\pi/4 Re \cdot Pr\, d/z > 10$.

B. Turbulent Flow

It is necessary to distinguish between completely developed and perturbed velocity profiles, among liquid, vapor/gases, and liquid metal flows, and between circular and noncircular ducts.

For the evaluation of the heat transfer coefficient for a flow with a completely developed velocity profile inside circular ducts, the most known and used relationship is[3]

$$Nu = 0.023\, Re^{0.8}\, Pr^n \tag{13}$$

$n = 0.3$, cooling of the fluid (liquid); $n = 0.4$, heating of the fluid (liquid). Equation 13 is also applicable to gases at low ΔT. With gases, it generally underpredicts the values of measured heat transfer coefficients. Equation 13 is applicable in the following ranges: $10^4 < Re < 1.2 \cdot 10^5$, $0.7 \leqslant Pr \leqslant 120$, and $L/d > 60$ and is quite easy to use, but it is approximate. The exponents of Re and Pr and the coefficient should vary with the variations of the fluid physical properties and with the temperature.

If this effect is important for the specific application, a more adequate correlation structure is needed[4]:

$$Nu = 5 + 0.015\, Re^a\, Pr^b \tag{14}$$

where:

$$a = 0.88 - \frac{0.24}{4 + Pr}$$

$$b = 0.33 + 0.5 \exp(-0.6\,Pr)$$

valid in the following ranges: $10^4 \leqslant Re \leqslant 10^6$, $0.1 \leqslant Pr \leqslant 10^4$, and $L/d > 60$.

Another method, less accurate but simpler, suggested by many authors, is that of modifying the Nu value evaluated through the assumption of constant properties of the fluid, using the factor:

$$\left(\frac{\mu_b}{\mu_w}\right)^a$$

$a = 0.11$ (heated fluid); $a = 0.25$ (cooled fluid), with the limits:

$$10^4 \leqslant Re \leqslant 1.25 \cdot 10^5; \quad 2 \leqslant Pr \leqslant 140; \quad 0.08 \leqslant \frac{\mu_w}{\mu_b} < 40$$

A comparison between Equations 13 and 14 in the range $1 \cdot 10^4 \leqslant Re \leqslant 2 \cdot 10^5$ and $0.5 \leqslant Pr \leqslant 1.5$ shows a generally good agreement, with a tendency of Equation 14 to prevail on Equation 13 with an increasing Prandtl number.

In single-phase flows of gas (or vapor) at high pressure and temperature, the following relationship is proposed[5]:

$$Nu = 0.0126\, Re_f^{0.89}\, Pr_f^{0.33} \left(\frac{L}{d}\right)^{-0.13} \tag{15}$$

obtained with experimental data in the following ranges:

$$d \cong 0.0109 \text{ ft}; \leftrightarrow 14.7 < \frac{L}{d} < 80; \leftrightarrow 115 < p < 3500 \text{ psia};$$

$$55{,}000 < G < 165{,}000 \frac{lb}{ft^2\, hr}; \leftrightarrow 607 < T_w < 1288°F;$$

$$429 < T_b < 1012°F; \leftrightarrow 91 < h < 354 \frac{Btu}{hr\, ft^2\, °F}$$

With specific reference to saturated and superheated steam in the ranges of interest in nuclear steam generators, it is suggested[6]:

$$Nu = 0.0142\, Re^{0.824}\, Pr^{0.98} \tag{16}$$

with the applicability field:

$$60 < p < 160 \frac{kg}{cm^2}; \leftrightarrow 100 < G < 250 \frac{g}{cm^2\, 5}; \leftrightarrow T_b < 400°C;$$

$$T_w < 450°C; \leftrightarrow 30 < q'' < 60 \frac{W}{cm^2}$$

This correlation has the advantage, with respect to others proposed in literature,[7,8] of being particularly simple because the physical properties are referred to the bulk temperature.

In the inlet zone of the tubes, where the velocity profile is not completely developed and the turbulence level near the wall is greater, an increase in heat transfer coefficients occurs. Such an effect is often rather important in gas flows.

For the local heat transfer coefficient in the first part of a tube (L/d < 60) and with reference to a water flow, the following relationship is proposed:

$$Nu(z) = Nu_o \cdot \left(\frac{d}{z}\right)^{0.054} \cdot 1.57 \tag{17}$$

where z is the distance from the inlet section and Nu_o is the value obtained in absence of intensification, making use of Equation 13. The applicability ranges are

$$Re > 8 \cdot 10^3, \leftrightarrow \frac{L}{d} < 60$$

With reference to gas flows, the entrance effect may be evaluated in function of the Reynolds number using the following relationship:

$$Nu(z) = Nu_o \cdot \epsilon \tag{18}$$

$$\epsilon = 1 + 0.48\left(1 + \frac{3600}{Re\sqrt{\frac{z}{d}}}\right) \cdot \exp\left(-0.17\frac{z}{d}\right) \cdot \left(\frac{z}{d}\right)^{-0.25} \tag{19}$$

where Nu_o is evaluated through Equation 13.

The bend region of tubes has been studied in Reference 23, where relevant perturbations in water downstream bends of 180° are pointed out. In order to quantify the bend influence in the evaluation of the heat transfer coefficient, the following criteria are suggested:

1. When Re ≤10,000, the effect of bends may be neglected and the correlations to be used are the same proposed for straight tubes
2. When Re >10,000, the effect of bends is not negligible, and the heat transfer coefficient evaluated with reference to straight tubes must be corrected using a coefficient, C, which assumes the following values:

   $$\begin{aligned} C &= 1.25 && \text{if } R_c/r < 4 \\ C &= 1.175 && \text{if } R_c/r = 4\text{—}8 \\ C &= 1.1 && \text{if } R_c/r = 8\text{—}25 \\ C &= 1 && \text{if } R_c/r \geq 25 \end{aligned}$$

   where R_c is the channel curvature radius and r is its internal radius.

The experimental validity range is $7 \cdot 10^2 \leq Re \leq 2 \cdot 10^4$.

A method often used to increase heat transfer coefficients inside tubes without increasing the average fluid velocity (to avoid excessive pressure drops) is that of creating, inside the tubes, conditions for the destruction of the laminar sublayer, in proximity to the wall. This method may be obtained through turbulence promoters, which are interesting mainly in gas or vapor flows.

The effect of marked superficial roughnesses, of internal finning, or of other turbulence promoters has been pointed out in several analyses, aimed both at the comprehension of the local phenomenology and at the definition of the optimal geometric characteristics of the turbulence promoter.

Particular importance has been attributed to the promoters of the twisted-tape (or twisted-wire) kind, mainly due to their simplicity of installation. The existence of an optimal disposition pitch, P, of the tape has been recognized (around values of $P/d \simeq 10$), also considering the friction losses effect, while the influence of the thickness of the twisted tape does not seem relevant. The effect of internal wiring in a tube with $d = 1$ cm crossed by air is reported in Reference 22.

The tube (pipe) circular section is undoubtedly the most used. Rectangular or square duct sections are also used, especially for gases. Many experiments have been performed in order to evaluate the influence, on the heat transfer coefficient, of such a geometrical parameter (section shape).

If the flowing liquid is a liquid metal, the well-known correlation of Martinelli-Lyon is suggested, which is applicable to sodium and sodium-potassium mixtures:

$$Nu = 7 + 0.025(Re \cdot Pr)^{0.8} \tag{20}$$

Its validity has been verified with channels subjected to a constant heat flux and with Pe >100.

In the same range, but hypothesizing a constant wall temperature, T_w, in Reference 9 the following relationship is proposed:

$$Nu = 5 + 0.025(Re \cdot Pr)^{0.8} \tag{21}$$

Equation 21 may be used only if the liquid metal is very pure. The metal purity is a highly influencing parameter and requires great care in defining and applying heat transfer correlations.

In Reference 10, a relationship is proposed, applicable to liquid metals in steel tubes without particular limits in the impurity content (oxygen content):

$$Nu = 3 + 0.014(Re \cdot Pr)^{0.8} \tag{22}$$

Equation 22 gives heat transfer coefficients which are conservative, in the sense that they underpredict the really measured values.

With coiled layout of tubes (LMFBR heat exchangers), in order to evaluate the heat transfer coefficient for the fluid flowing inside them, the same formulas proposed for straight tubes may be used, multiplying the results by a factor between 1.1 and 1.2.

III. FLOW OUTSIDE TUBES

A great number of studies have been dedicated recently to the heat transfer analysis of fluids flowing outside tubes, in parallel or in cross directions. The efficiency of each tube row in a bundle and the overall efficiency of the bundle itself with various heat transfer typologies have been analyzed. The influence of the longitudinal and of the transversal pitch has also been studied, together with the effects of voids in the layout of the tubes and the performances of bundles crossed by flows with a parallel and an orthogonal direction with respect to their axis, have been systematically compared.

In the usual ranges of the pitch/diameter (P/d_o) ratio, heat transfer in cross flow with staggered tubes has been verified as the most effective, with respect to an in-line tube disposition or with respect to a parallel flow. Pressure drops are generally higher in the

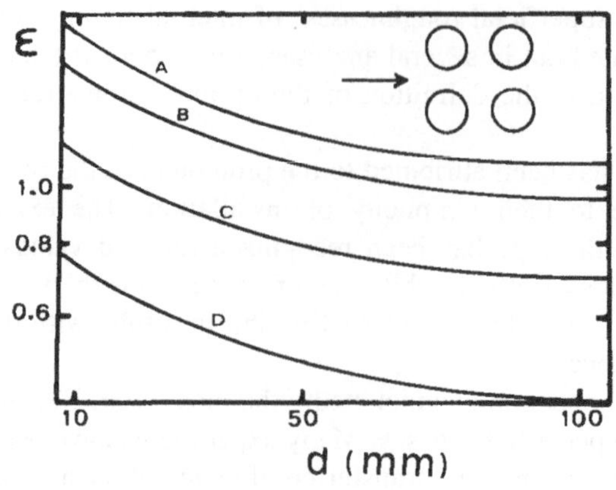

FIGURE 1. Efficiency as a function of tube diameter. (A) 1.10 × 1.10; (B) 1.25 × 1.25; (C) 1.50 × 1.50; (D) 2.00 × 2.00).

configurations which are characterized by higher heat transfer coefficients. Furthermore, the relative efficiency of the different rows of the tubes in a bundle with a cross flow is a function of geometry and the turbulence level.

With in-line tubes in a cross flow, with a geometry 2 × 2* and $1 \leq Re \leq 1000$, the heat transfer coefficient of the first row is greater than that of an internal row, of a factor 1.25.[11,12] With a geometry 1.5 × 1.5, the greater efficiency of the first row is about 10%; with staggered tubes and cross flows, a wide variation interval of heat transfer coefficients has been verified (from 3 to 40%), with the variation of the ratio (P/d_o) and of the Reynolds number. Furthermore, with bundles subjected to a cross flow, a dependence of the parameter efficiency** on the tube diameter, has been found (Figure 1).

In order to characterize the performance of each configuration, it is necessary, first of all, to distinguish between cross and parallel flow.

A. Cross Flow

For the evaluation of the heat transfer coefficient, in Reference 13 the following definitions are adopted:

1. Flow Reynolds number: $Re = G\,d/\mu$, where G is the mass velocity, evaluated with reference to the minimum flow section
2. Fraction of area occupied by the fluid (see Figure 2): $E = (A_{tot} - A_{tubes})/A_{tot}$, where A_{tot} is the overall cross area of the bundle and A_{tubes} is the area occupied by the tubes.

The following correlation is then proposed, which has been confirmed by several independent tests:

$$J\,E^{\phi b} = a\left(\frac{Re}{E}\right)^{c} \tag{23}$$

$$b = 0.175\left(\frac{Re}{E}\right)^{-0.07} \tag{24}$$

* (transverse pitch/diameter) × (longitudinal pitch/diameter).
** Ratio between heat exchanged and pressure drops.

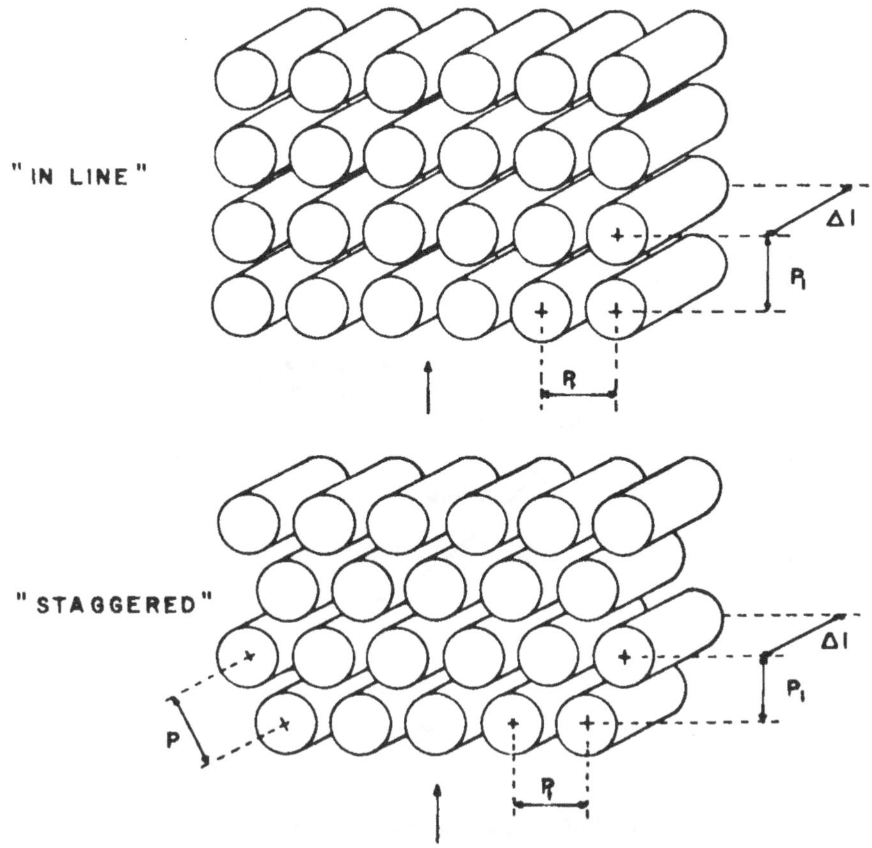

FIGURE 2. Typical cross flow tube array.

Table 1
PARAMETERS a AND c IN EQUATION 23

Staggered Array

Re/E	1—100	300—40,000	>80,000
a	0.895	0.38	0.051
c	−0.59	−0.39	−0.2

In-Line Array

Re/E	1—300	2,000—30,000
a	0.785	0.101
c	−0.62	−0.255

where a and c are in Table 1 and ϕ may be obtained from Figure 3 as a function of the ratio P_t/P_l, and

$$J = \left(\frac{h}{c_p\,G}\right)\cdot\left(\frac{c_p\,\mu}{k}\right)^{2/3}\cdot\left(\frac{\mu_b}{\mu_w}\right)^{0.14} \quad (25)$$

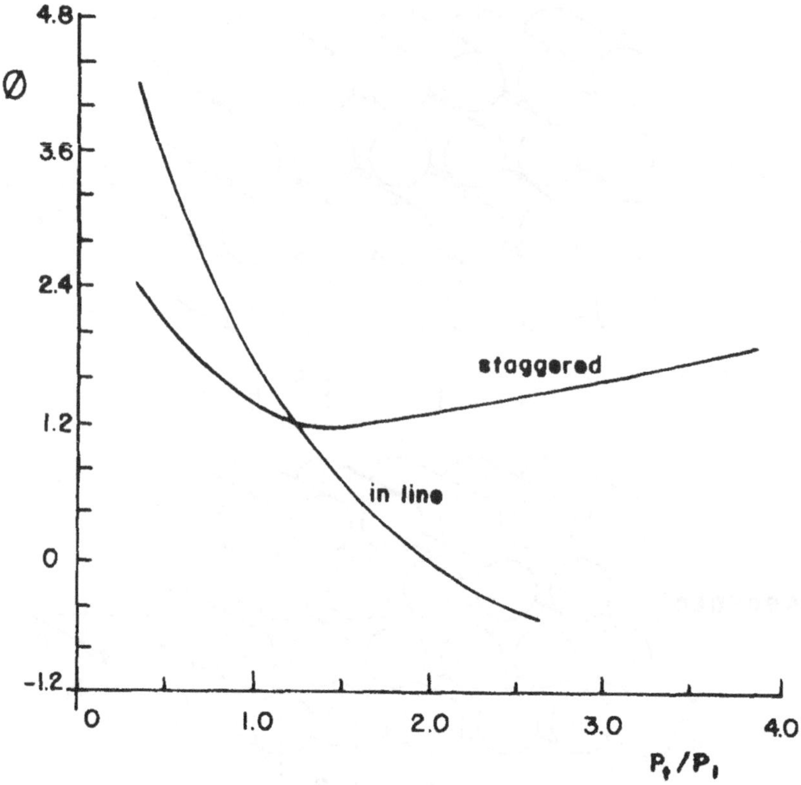

FIGURE 3. Factor ϕ of Equation 23.

Recently, the influence of blockages and disturbances of the flow on the performance of tube bundles has been subjected to investigation. Until now, definitive results have not been available, owing mainly to the numerous effects to be simulated.

For a tube bundle with a blockage of 50% of the inlet section, making use of the analogy between mass and heat transfer, maps as in Figure 4 are proposed in Reference 4. They represent the variations of the Sherwood number with respect to undisturbed bundle conditions. Such variations are then correlated to the heat transfer coefficient, making use of the Reynolds analogy, from which:

$$Nu = Sh(Pr/Sc)^n \tag{26}$$

where Pr is evaluated with air in normal conditions and Sc $= D/\bar{\nu}$, D being the mass diffusion coefficient of the test substances (naphthalene in air).

B. Parallel Flow

If a single-phase fluid flows through a tube bundle parallel to its axis, its heat transfer coefficient may be evaluated (in the undisturbed flow zones) using the structure of the Dittus-Boelter relationship:

$$Nu = a \cdot Re^b \cdot Pr^c \tag{27}$$

where factor a is usually expressed as a function of the bundle geometry, while the exponents b and c are analogous to those used for the flow inside tubes (see Equation 13). For the evaluation of factor a, in Reference 15 it is proposed:

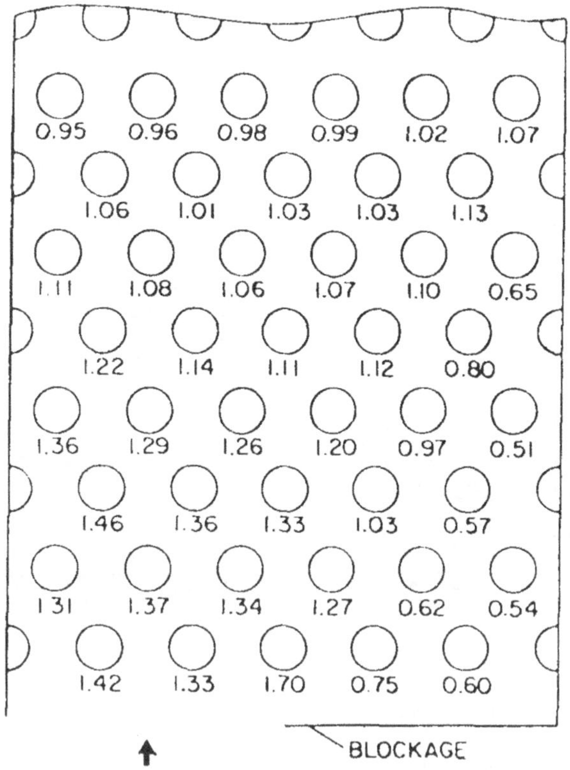

FIGURE 4. Per tube ratio of the maldistribution-affected Sherwood number to that for uniform inlet flow, Re = 8400.

1. Triangular pitch lattice: $a = 0.026\ (P/d_o) - 0.006$
2. Square pitch lattice: $a = 0.042\ (P/d_o) - 0.024$

with $Nu = h\ d_{eq}/k$, with $d_{eq} = 4\ A/P_h$, equivalent diameter* and P_h, heated perimeter.

If a liquid metal (sodium) is flowing inside a tube bundle with a ratio $P/d_o \geq 1.35$, in the absence of turbulence promoters, the following relationship may be used:[24]

$$Nu = \frac{h\ d}{k} = \alpha + \beta(\bar{\psi}\ Pe)^{\gamma}; \leftrightarrow 10^2 < Pe < 10^4 \tag{28}$$

being

$$\bar{\psi} = 1 - 1.82/[Pr(\nu_t/\nu)_{max}]^{1.4}$$

The constants α, β, and γ are evaluated through the following:

* The use of the concept of the equivalent diameter is not limited to the pressure drops evaluation (see Chapter 4), but is extended to that of heat transfer coefficient in flow geometries different from that of a circular tube. Validity limits of such use are underlined by the following experimental results obtained with fluids (air and water) in parallel flow in rod bundles: (1) for $P/d_o = 1$ (rods in contact), pressure drops have resulted in about 65% of those evaluated for tubes with the same equivalent diameter; (2) for $P/d_o = 1.15 - 1.25$, the measured pressure drops resulted greater than those evaluated for the equivalent tube, with a factor of 1.1 to 1.2. For heat transfer calculation purposes, the use of the equivalent diameter concept seems to lead to quite good results.

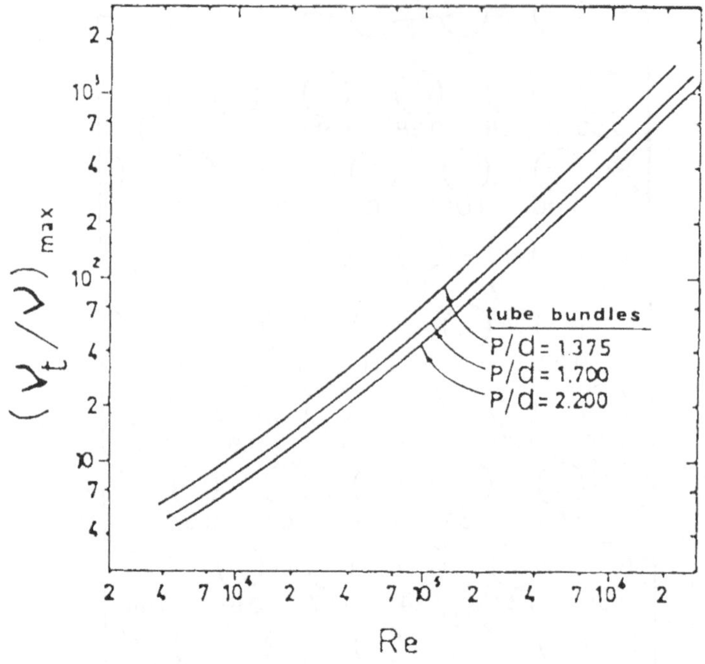

FIGURE 5. Values of (ν_t/ν_{max}) vs. Re for Equation 28.

- Triangular lattice $(1.3 \leq P/d \leq 3)$:

$$\alpha = 6.66 + 3.126(P/d_o) + 1.184(P/d_o)^2; \quad \beta = 0.0155; \quad \gamma = 0.86$$

- Square lattice $(1.7 \leq P/d_o \leq 10)$:

$$\alpha = 7 + 4.24(P/d_o)^{1.52}; \quad \beta = 0.0275(P/d_o)^{0.27}; \quad \gamma = 0.8$$

The parameter (ν_t/ν_{max}) is evaluated through Figure 5. In many applications of liquid metals in parallel flow, use is made of vorticators, which cause an enhanced mixing in the sodium flow (see Figure 6).

The presence of grids and spacers is also of interest in the situation of single-phase flows of gas (vapor) or liquid inside rod bundles. Such components break the laminar sublayer, thus determining an increase in the heat transfer coefficient. Specific measurements are suggested, with reference to the particular geometry.

With specific reference to the grids used in nuclear industry and with a vapor flow (postburnout accidental conditions in light water reactor fuel bundles), in Reference 16 the following relationship, deduced from Reference 17, is proposed:

$$\frac{Nu}{Nu_o} = 1 + 5.55 \cdot \epsilon^2 \cdot \exp[-0.13(z/d_o)] \tag{29}$$

where ϵ = blockage ratio of the grid (ratio between the free flowing area in correspondence to the grid and the flowing area in absence of the grid), z = distance from the end of the grid, and Nu_o = local Nu at the wall without the grid installed. The verified application range is $1.5 \cdot 10^4 \leq Re \leq 15.5 \cdot 10^4$ and $0.15 \leq \epsilon \leq 0.36$.

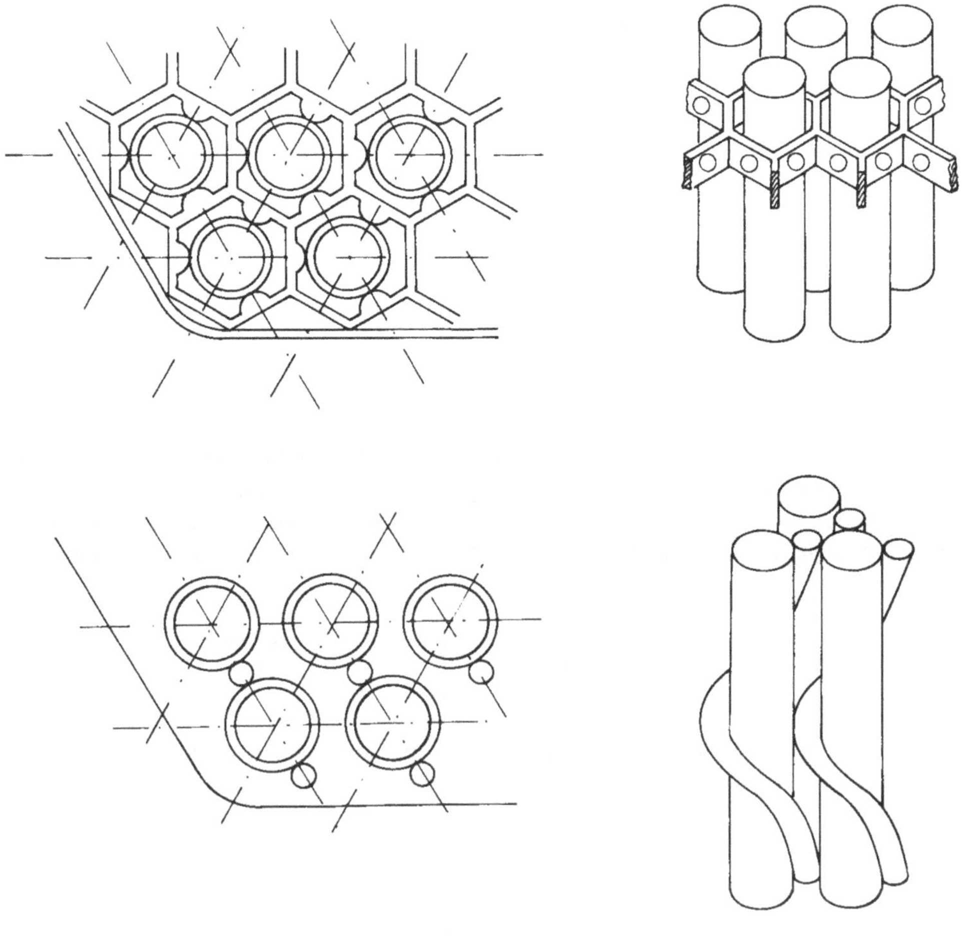

FIGURE 6. Typical spacers/vorticators configurations.

C. Finned Tube Bundles

In heat transfer equipment, the technique of increasing (gas or vapor side) the contact surface by means of fins is often used as a consequence of the low value of heat transfer coefficients obtainable with gas or vapor flows. This technique is mainly used with cross flows.

Many types of fins are used (radial/longitudinal; spike/continuous; etc.); the maximum of the fin efficiency is obtained when a uniform "flow area" is offered to the thermal flux to/from the tube. The fin type must be carefully selected owing to possible problems in the equipment cleaning. Such solutions allow very high efficiency and a savings in the added material of the order of 50 to 80% (in function, also, of the thermal conduction of the fin and the layout of tubes). The effect of a fin is not only an increase of the contact surface between the fluid and the wall, but also the destruction of the viscous boundary layer in contact with the surface, inducing a local turbulence augmentation.

The most common fin surface configurations are characterized by a fin density of 1 to 10 fins per centimeter (1 to 26 fins per inch), with fins in aluminum or steel, on copper or steel tubes.

For the evaluation of the heat transfer coefficient in tube bundles with a cross flow (especially aimed at MSRs), a correlation is proposed in Reference 18 (see Table 1 of Chapter 4):

$$J_h = 0.292\left(\frac{s}{d_f}\right)^{1.115} \cdot \left(\frac{s}{l}\right)^{0.257} \cdot \left(\frac{B}{s}\right)^{0.666} \cdot \left(\frac{d_f}{d_r}\right)^{0.473}$$

$$\cdot \left(\frac{d_f}{B}\right)^{0.7717} \cdot Re^{[-0.415+0.0346\ \ln(d_f/s)]}$$

$$J_h = \left(\frac{h \cdot Pr^{2/3}}{c_p\ G_{max}}\right) \qquad (30)$$

with

$$Re = \frac{d_r\ G_{max}}{\mu}$$

Owing to the great influence of supporting systems in bundles on heat transfer behavior, in Reference 19, the importance of leakage and bypass effects is analyzed.

For the evaluation of the heat transfer coefficient in tube bundles with a parallel flow, little information is available in the literature; the validity of the following correlation is confirmed in Reference 20:

$$Nu = 0.037(Re^{0.75} - 180) \cdot Pr^{0.42} \cdot \left[1 + \left(\frac{d_{eq}}{L}\right)^{2/3}\right]\left(\frac{Pr}{Pr_w}\right)^{0.14} \qquad (31)$$

which was proposed in Reference 21, where d_{eq} is the equivalent diameter and L is the bundle length.

REFERENCES

1. **Cumo, M., Lopez, S., and Pinchera, G. C.,** Effetto di imbocco: calcolo dello strato limite fluidodinamico, paper presented at ATI Annu. Meet., Siena, 1964.
2. **Cumo, M., Farello, G. E., and Pinchera, G. C.,** Determinazioni sperimentali della lunghezza di assetto fluidodinamico in canali a sezione circolare, paper presented at ATI Annu. Meet., Genova, 1965.
3. **McAdams, W.,** *Heat Transmission,* McGraw Hill, New York, 1954.
4. **Notter, R. H. and Sleicher, C. A.,** *Chem. Eng. Sci.,* 27, 1972.
5. **McAdams, W., Kennel, W. E., and Addonis, J. N.,** Heat transfer to superheated steam at high pressures, *Trans. ASME,* 72, May 1950.
6. **Campolunghi, F., Cumo, M., Ferrari, G., and Vaccaro, G.,** Heat transfer to dry and superheated steam in once-through generators for LMFBR CNEN RT/ING, 74, 18, 1974.
7. **Heinemann, J. B.,** An experimental investigation of heat transfer to superheated steam in round and rectangular channels, *ANL,* 6213, 1966.
8. **Miropolski, Z. L.,** Heat Transfer in Steam Generating Tubes, AEC TR 6252 Teploenergetica, 10(5), 1963.
9. **Seban, R. A. and Shimazaki, T. J.,** Heat transfer to a fluid flowing turbolently in smooth pipes with walls at constant temperature, *Trans. ASME,* 73, 1951.
10. **Kutateladze, S. S.,** *Fundamentals in Heat Transfer,* Pergamon Press, Oxford, 1966.
11. **Zhukauskas, A. A. and Ulinskas, R. V.,** Heat transfer efficiency of tube bundles in cross flow at critical Reynolds numbers, *Heat Transfer-Sov. Res.,* 10(5), September/October, 1978.
12. **Zhukauskas, A. A., Ulinskas, R. V., and Sipavicius, C. S. J.,** Average heat transfer and pressure drop in cross flow of viscous fluids over tube bundles at low Re, *Heat Transfer-Sov. Res.,* 10(6), November/December 1978.
13. **Weisman, J.,** Effect of void volume and Prandtl modulus on heat transfer in tube banks and packed beds, *AIChE J.,* September 1955.
14. **Sparrow, E. M. and Ruiz, R.,** Effect of blockage — introduced flow maldistribution on the heat transfer and pressure drop in a tube bank, *J. Heat Transfer,* 104, November 1982.

15. **Tong, L. S. and Weisman, J.,** *PWR's Thermal Analysis,* 2nd ed., American Nuclear Society, LaGrange, Ill., 1979.
16. **Yao, S. C., Hochreiter, L. E., and Leech, W. J.,** Heat transfer argumentation in rod bundles near grid spacers, *J. Heat Transfer,* 104, February 1982.
17. **Marek, J. and Rehme, K.,** Heat transfer in smooth and roughned rod bundles near spacer grids, in *Fluid Flow and Heat Transfer Over Rod or Tube Bundles,* Yao, S. and Pfund, P., Eds., American Society of Mechanical Engineers, New York, 1979.
18. **Rabas, T. J., Eckels, P. W., and Sabotino, R. A.,** The effect of fin density on the heat transfer and pressure drop performance of low-finned tube banks, paper presented at AIChE Natl., *Heat Transfer Conf.*, Orlando, Fla., July 27, 1980.
19. **Eckels, P. W. and Rabas, T. J.,** A finned tube heat exchanger side seal study, American Society of Mechanical Engineers, New York, 1981.
20. **Obermeier, E. and Schaber, A.,** Experimental investigation of heat transfer from transverse finned tubes with longitudinal flow, paper presented at 6th Int. Heat Transfer Conf., Toronto, August 1978.
21. **Hansen, H.,** *All. Waermetech.*, 9, 75, 1959.
22. **Sams, E. W.,** Heat transfer and pressure drops characteristics of wire-coil-type turbulence promoters, TID-7529, Part 1, Book, 2, 1957.
23. **Ede, E. J.,** Int. Dev. in Heat Transfer, Part 3, American Society of Mechanical Engineers, New York, 1961.
24. **Dwyer, O. E.,** Heat Transfer to Liquid Metals Flowing in Line Through Rod Bundles, American Society of Mechanical Engineers, New York, 1968.

INDEX

A

B

C

D

E

F

O

P

R

S

T

V

W

Z

O

P

R

S

T

V

W

Z